CARPENTRY AND BUILDING CONSTRUCTION STUDENT WORKBOOK

Fifth Edition

by John L. Feirer

Keyed to the fourth and fifth editions of the textbook
Carpentry and Building Construction

 **Glencoe McGraw-Hill**

New York, New York
Columbus, Ohio
Woodland Hills, California
Peoria, Illinois

Glencoe/McGraw-Hill

A Division of The McGraw·Hill Companies

Copyright © 1997 by Glencoe/McGraw-Hill. Previous copyright © 1993 by the Glencoe
Division of Macmillan/McGraw-Hill School Publishing Company. All rights reserved. Except
as permitted under the United States Copyright Act, no part of this publication may be
reproduced or distributed in any form or by any means, or stored in a database or retrieval
system, without the prior written permission of the publisher.

Send all inquiries to:
Glencoe/McGraw-Hill
3008 W. Willow Knolls Drive
Peoria, IL 61614

ISBN 0-02-838701-5

Printed in the United States of America

4 5 6 7 8 9 10 11 12   024   02 01 00 99

TABLE OF CONTENTS

TO THE TEACHER

This student workbook has been designed for use with the textbook *Carpentry and Building Construction*. It has been planned as a teaching and learning aid. The exercises it presents are meant to reinforce and evaluate learning.

This student workbook can be used by students individually or in group activity. Pages are perforated, allowing individual study units to be removed for ease in use and grading.

Students can be asked to complete the exercises in this student workbook in one of several ways:

1. Each student may progress at his or her individual rate of speed, completing the study unit exercises as quickly as possible.

2. The teacher may assign a unit or section of a unit in the textbook as homework. The appropriate study unit may then be assigned the next day to evaluate learning.

3. Students can be assigned sections of the textbook for outside study. They may then be asked, at the completion of their study assignment, to complete the appropriate study units in the student workbook.

A list of the possible scores in study unit exercises is included. This list will enable students to track their progress.

Note that the questions in each study unit in this workbook relate to a particular text unit or units. The unit or units to which the study unit refers is given at the top left-hand corner of the first page of each study unit. The textbook page numbers of the unit or units are given immediately below this text unit reference. Note that the student work study units do not carry the same numbers as the textbook units.

Other material has been included in this student workbook. This additional material will simplify the teaching and be of value in:

• Organizing the class.

• Keeping accurate records of the cost of carpentry materials.

• Maintaining an accurate safety record.

Space has been provided to write in additional regulations.

TO THE STUDENT
How to Use This Student Workbook

This student workbook includes the material you need for keeping a record of your progess in carpentry.

Complete the following:

• Fill in the *Student Information* sheet in this workbook. This will help your teacher become better acquainted with you.

• Keep a record of your clean-up assignments in the space provided in this workbook.

• Make a list of the regulations that you are expected to follow.

• Read the *Safety Pledge* carefully. Sign it after you have agreed to obey the safety regulations.

• Use the *Accident Report* sheet to record needed information in case of an accident.

• Fill in the *Cost of Materials* chart. These costs will be used when making the bill of materials for each project.

• Make a plan for each project you build.

• Fill in the information at the top of each study guide. The textbook pages you should study are given at the top left side of the first page of each study guide. Write the answers in the blank spaces along the left side of each page.

You will find that this student workbook has six different kinds of questions. Samples of each kind of question are given on the next page. Study each type of question so you will know the correct way to answer each question.

Sample Questions

_____ T _____

1. TRUE-FALSE (T-F). Read the statement carefully and decide if it is true or false. If the statement is true, put a T in the blank; if it is false, put an F.

   SAMPLE: Washington, D.C. is the capital of the United States. (T or F)

_____ C _____

2. MULTIPLE CHOICE WITH ONE RIGHT. Read the question and the possible answers. Select the best or correct answer and put the letter for this in the blank.

   SAMPLE: The following state is east of the Mississippi River (one right): a. Arkansas; b. Missouri; c. Illinois; d. Iowa.

_____ C _____

3. MULTIPLE CHOICE WITH ONE WRONG. Study the question and the possible answers. Select the one that is wrong and put the letter for this in the blank.

   SAMPLE: The following are common fractions (one wrong): a. 1/2; b. 3/8; c. 2; d. 3/4.

_____ blue _____
_____ east _____
_____ chicken _____

4. COMPLETION. Study the sentence and decide what word will best finish the sentence. Write this in the space at the left.

   SAMPLE: The primary colors are red, yellow, and _____.

   SAMPLE: The sun rises in the _____ and sets in the west.

   SAMPLE: Milk is to cow as egg is to _____.

5. MATCHING. Match each item in the left-hand column with the correct one in the right-hand column. Show your choice by placing the correct numbers in the spaces at the left.

   SAMPLE: Match the items at the left to the items at the right:

a. _____ 1 _____   a. goal posts   1. football
b. _____ 4 _____   b. hoop   2. tennis
c. _____ 3 _____   c. puck   3. hockey
d. _____ 2 _____   d. racquet   4. basketball

a. _____ knife _____

b. _____ fork _____

c. _____ spoon _____

6. NAME OR IDENTIFICATION. Write in the space at the left the correct name of the item shown.

   SAMPLE: Name these common eating utensils:

a   b

c

STUDENT INFORMATION

Print:

1. Name _____
                          Last   First   Middle

2. Home address _____

3. Home phone number _____

4. Year _____

5. School last attended _____

6. Parents' names:

   Father _____

   Mother _____

7. Parents' occupations:

   Father employed by _____

   Mother employed by _____

8. Hobbies or outside interests _____

_____

9. Previous carpentry experience _____

10. Name of family doctor _____

            Phone _____

       Address _____

CLEAN-UP ASSIGNMENTS

	Date	**Job**
From _____ to _____	_____
From _____ to _____	_____
From _____ to _____	_____

SAFETY REGULATIONS

1. _____

2. _____

3. _____

4. _____

5. _____

6. _____

7. _____

8. _____

9. _____

10. _____

SAFETY PLEDGE

I pledge that I will follow all of the safety rules given in the book *Carpentry and Building Construction* and all of the regulations listed above. I will not use a power tool without first securing the permission of the instructor. I will report all accidents to the instructor immediately, no matter how small they are. I will help to maintain a safe environment by tending to my business and by not bothering other students who are busy.

Name _____ Date _____

ACCIDENT REPORT

1. Name of injured _____

    Address _____

    Telephone _____

2. Nature of injury (Cut, scratch, foreign matter in eye, etc.)

    _____

3. Tools or machines being used _____

4. Witnesses to the accident: Name _____

    Address _____

    Name _____

    Address _____

5. Treatment: First aid _____ By whom _____

    Physician _____ Address _____

    Hospital _____ Address _____

6. Cause of accident (Poor condition, wrong procedure, etc.)

    _____

    _____

7. Correction (What will be done to prevent future accidents)

    _____

    _____

    _____
    Name

COST OF MATERIALS

Lumber:

Kind	Cost Per Board Foot	Kind	Cost Per Board Foot
Pine			
Fir			
Maple			
Oak			
Birch			

Panel Stock:

Kind	Grade	Thickness	Cost Per Sq. Foot
Plywood			
Hardboard			
Particleboard			

x

COST OF MATERIALS (cont.)

Screws, Nails, and Other Items:

No. or Size	Kind	Cost Per _____
	Screws	
_____	_____	_____
_____	_____	_____
	Nails	
_____	_____	_____
_____	_____	_____
	Dowel	
_____	_____	_____
	Sandpaper	
_____	_____	_____
_____	_____	_____
	Hardware	
_____	_____	_____
_____	_____	_____
_____	_____	_____

Finishing Materials:

Kind	Cost Per _____
_____	_____
_____	_____
_____	_____
_____	_____
_____	_____

STUDY UNIT SCORES

STUDY UNIT NUMBER	POSSIBLE SCORE	NUMBER CORRECT	STUDY UNIT NUMBER	POSSIBLE SCORE	NUMBER CORRECT
1	42	_____	31	35	_____
2	46	_____	32	25	_____
3	38	_____	33	34	_____
4	59	_____	34	44	_____
5	52	_____	35	34	_____
6	42	_____	36	41	_____
7	47	_____	37	57	_____
8	169	_____	38	24	_____
9	32	_____	39	26	_____
10	60	_____	40	30	_____
11	41	_____	41	38	_____
12	29	_____	42	62	_____
13	35	_____	43	38	_____
14	61	_____	44	21	_____
15	51	_____	45	30	_____
16	28	_____	46	21	_____
17	32	_____	47	42	_____
18	53	_____	48	64	_____
19	43	_____	49	32	_____
20	29	_____	50	38	_____
21	25	_____	51	41	_____
22	33	_____	52	42	_____
23	33	_____	53	46	_____
24	42	_____	54	27	_____
25	43	_____	55	65	_____
26	28	_____	56	49	_____
27	39	_____	57	23	_____
28	43	_____	58	21	_____
29	33	_____	59	33	_____
30	23	_____	60	28	_____

Name _____

Score: (42 possible) _____

Study Unit 1
Careers and Planning

_____T_____

_____T_____

_____T_____

_____Technical_____

_____b_____

_____F_____

_____journeyman_____

_____T_____

_____T____1,000,000_____

_____F_____

_____T_____

_____T_____

_____An apprentice_____

_____technical_____

_____F_____

_____2½_____

1. One of the largest industries in the United States is construction. (T or F)

2. An example of light construction would be home building. (T or F)

3. Some home building today is done with prefabricated components. (T or F)

4. Three career levels that require special training and education are craft, _____, and professional.

5. Examples of craft careers in building construction are *(one wrong)*: a. carpenter; b. architect; c. plumber; d. painter.

6. There are fewer than one dozen skilled building trades. (T or F)

7. A _____ is a craft worker in unions.

8. According to Table 1-A, the growth in the number of manufacturing jobs is greater than the number of construction jobs. (T or F)

9. There are over 800,000 carpenters employed in the construction trades. (T or F)

10. The percentage change in employment in construction from 1988-2000 will be approximately 25%. (T or F)

11. Specialization is common in carpenters' trades. (T or F)

12. It is possible for a carpenter to learn the trade without formal education. (T or F)

13. One who is entering construction by combining on-the-job training with basic instruction is called a _____.

14. Courses like drafting, building techniques, and surveying are examples of preparation for _____ careers in building construction.

15. A college degree is required for professional jobs in building construction. (T or F)

16. Housing for a family should not cost more than _____ times the average annual income of the family.

17. Consider the following when choosing a lot for a house:

a. _____Sun_____

b. _____Recreational_____

c. _____deep/wide_____

d. _____Stable_____

a. Is it possible to use the _____ to supply part of heating and electrical needs?

b. Are jobs, schools, community services and _____ facilities nearby?

c. Is the lot shallow or _____?

d. Is the neighborhood likely to remain relatively _____?

(Continued on next page)

_____ utilities _____

a. _____ survey _____
b. _____ deed _____
c. _____ abstract of title _____

_____ T _____

_____ T _____

_____ 5-10% _____

_____ T _____

_____ bid _____

_____ lien _____

_____ 4D _____

_____ Closing _____

_____ escrow _____

_____ % _____

_____ safety _____ ✓

a. _____ Uniform _____ ✓
b. _____ Basic building _____ ✓
c. _____ Standard _____ ✓
_____ National Electric Code _____ ✓

_____ Permit _____ ✓

_____ inspector _____ ✓

_____ Certificate _____ ✓

18. Not more than 25% of monthly income should be spent for all housing expenses, including mortgage payments, _____, and repairs.

19. Three legal documents are needed before buying property:
   a. The _____ that shows the boundaries of the property.
   b. The _____ that is evidence of ownership.
   c. The _____ _____ _____ which is a history of deeds and other papers.

20. House plans can be purchased from companies that specialize in designing stock house plans. (T or F)

21. Purchasing stock plans is the least expensive way of obtaining house plans. (T or F)

22. If the architect only designs the house, the fee is usually _____ percent of the total building cost.

23. A rough estimate of how much a house should cost can be found by multiplying the average cost of building per square foot in your locality by the number of square feet in the house plans. (T or F)

24. To get an accurate price estimate of a home, ask a contractor to _____ on the home.

25. A mortgage loan is a _____ on the property.

26. Typical mortgages run from 12 to _____ years.

27. Charges for paperwork and similar items needed to make a loan are called _____ costs.

28. Parts of mortgage payment used to pay taxes and insurance are held in _____.

29. Contractors are usually paid a certain _____ of the construction costs before the home is started.

30. Building codes establish minimum standards of quality and _____ in housing.

31. The three major building codes in the United States include the:
   a. _____ Building Code used in the West.
   b. _____ Building Code used in the Midwest.
   c. _____ Building Code used in the East.

32. The electrical code used throughout the United States is the _____ _____ _____.

33. It is necessary to get a building _____ before construction can begin.

34. At key points during construction, a building _____ will visit the job site to examine the work.

35. If there are no problems after the house has been completed, a _____ of Occupancy will be issued.

Name _____

Score: (46 possible) _____

Study Unit 2
Reading Prints

1. Designers and architects express their ideas in drawings by means of lines, symbols, and _____.

2. The ability to read and understand plans, drawings, and _____ is basic to all construction.

3. An exact copy of a drawing is called a _____.

4. The building industry makes use of prints called _____.

5. Blueprints do not fade. (T or F)   *Prints fade not blue prints*

6. The metre is a metric symbol for _____.

7. A yard is a little shorter than a _____.

8. Metric units for length are *(one wrong)*: a. kilometre; b. millimetre; c. kilogram; d. centimetre.

9. One inch is equal to *(one right)*: a. 25.4 millimetres; b. 2.5 millimetres; c. 5.4 millimetres; d. 5.4 centimetres.

10. In building construction the inch is rounded off to _____ millimetres.

11. The litre is a measure for _____.

12. The litre is slightly larger than the _____.

13. The kilogram is a metric measure of _____.

14. A kilogram equals about *(one right)*: a. 1.8 pounds; b. 2.8 pounds; c. 2.2 pounds; d. 3.2 pounds.

15. In architectural drawings, metric measurements are given in _____.

16. Scale represents the ratio between the size of an object as drawn and its _____ size.

17. Scale is a unit of measurement. (T or F)

18. When a drawing is the same size as the object itself, it is called a _____ scale drawing.

19. A house plan that is drawn 3″ to 1′ is 1/4 full size. (T or F)   *most standard size 1/4″= 1′*

20. Figure 3-5 in the text shows two kinds of scales, the architects' scale and the _____ scale.

21. A perspective drawing, also called a _____ drawing, is a kind of pictorial drawing.

22. The kind of drawing prepared by the construction drafter is an architectural _____ drawing.

(Continued on next page)

Answers (handwritten):
1. dimensions
2. prints
3. print
4. blueprints
5. T
6. length
7. metre
8. C is wrong
9. a
10. 25
11. liquid
12. quart
13. weight
14. c
15. millimeters
16. actual
17. F
18. Full size or full scale
19. T
20. metric
21. presentation
22. working

a. _____Lines_____ ✓

b. _____dimension_____ ✓

c. _____Symbols_____ ✓

d. _____notes_____ ✓

a. _____tile_____

b. _____Sand_____

c. _____Cinder block_____

d. _____brick_____

e. _____Earth_____

f. _____Stone_____

23. Drawings are made of the following:

a. _____, which show the shape.

b. Numbers telling sizes, called _____.

c. Representations of things which would be difficult to draw, called

_____.

d. Written information and explanations, called _____.

24. Identify the symbols for the kinds of materials given in Fig. 2-1.

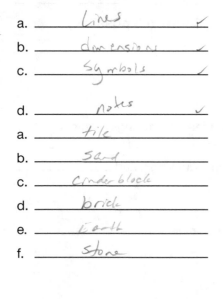

a   b   c

d   e   f

Fig. 2-1.

_____d is wrong_____ ✓

_____joists or trusses_____ ✓

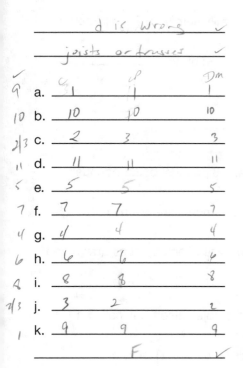

9 a. __1__   __1__   __1__
10 b. __10__   __10__   10
2/3 c. __2__   __3__   3
11 d. __11__   __11__   11
5 e. __5__   __5__   5
7 f. __7__   __7__   7
4 g. __4__   __4__   4
6 h. __6__   __6__   6
8 i. __8__   __8__   8
7/3 j. __3__   __2__   2
1 k. __9__   __9__   9

_____F_____ ✓

25. Parts of a roof are (one wrong): a. rafter; b. purlin; c. truss; d. girder. 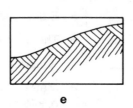 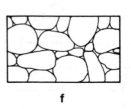 purline wall over 8' tall to keep studs from shifting

26. Subflooring is laid directly over _____.

27. Match the items on the left with their descriptions on the right:

a. specifications
b. sectional views
c. site plan
d. detail or detail drawings
e. building or floor plan
f. shop sketch
g. foundation or basement plan
h. elevations
i. framing plan
j. plot plan
k. bill of materials

1. table of information on needs for building a project
2. outline of lot
3. location of building on site
4. shows exact size of foundations and their location
5. cross-section view of house
6. show exterior view of house
7. simple drawing
8. shows structural members of house
9. written description of all materials
10. show the cross section of structural parts
11. show specific details of a house

28. All notes concerning a set of construction drawings should be placed on the drawings themselves. (T or F)

Study Unit 3
Energy Use in the Home

_____ ⅕ _____ ✓ 1. Energy consumed in living units is approximately one-_____ of our total national energy consumption.

_____ C is wrong _____ ✓ 2. The following are examples of exfiltration in the winter *(one wrong)*: a. leakage of warm air through cracks; b. leakage of warm air through windows; c. leakage of cold outside air into the house; d. leakage of warm air through doors.

_____ T _____ ✓ 3. Heat is lost through water drains in the living unit. (T or F)

_____ T _____ ✓ 4. Some heat is gained through solar energy, even in the standard house. (T or F)

_____ infiltration _____ ✓ 5. The heat that tends to flow into a structure is called _____.

_____ 40% _____ ✓ 6. Nearly _____ percent of the heat loss in homes is due to exfiltration and infiltration.

_____ humidity _____ ✓ 7. In cold weather, energy is needed to heat the air, clean it, and add _____ for comfort.

_____ F _____ ✓ 8. Dark-colored roofs reduce solar heat gain. (T or F)

_____ 15 _____ ✓ 9. The "payback" period for energy-saving devices is from one to _____ years.

_____ T _____ ✓ 10. An all-purpose living area is called a great room. (T or F)

_____ T _____ ✓ 11. Caulking is used to seal the house and prevent air leakage. (T or F)

_____ Cartridges _____ ✓ 12. Most caulks and sealants come in a sealed _____ for use in a caulking gun.

_____ F (45) _____ ✓ 13. The temperature of the material to be caulked should be at least 60 degrees F. (T or F)

_____ T _____ ✓ 14. The least expensive caulk is oil-based. (T or F)

_____ 20 _____ ✓ 15. The approximate life span of silicone used outdoors is _____ or more years.

_____ F _____ ✓ 16. Solvent acrylic is recommended for indoor use. (T or F)

_____ T _____ ✓ 17. Polysulfide caulk resists chemicals. (T or F)

_____ T should be F 18. All insulating sheathing is made from some type of plastic. (T or F)

_____ T _____ ✓ 19. Sheet metal let-in braces are used with nonstructural insulating sheathing. (T or F)

_____ 4 _____ ✓ 20. There are _____ basic kinds of insulating sheathing. (T or F)

_____ Expanded polystyrene _____ 21. An insulating sheathing similar to the plastic used to make coffee cups is _____.

(Continued on next page)

polyisocyanurate

Extruded polystrene × 22. An insulating panel that is faced on both sides with foil is ______________.

C is wrong ✓ 23. Energy efficiency in a standard house can be improved by *(one wrong)*: a. using the correct kind of sheathing; b. adding the proper insulation; c. increasing the square footing of the house; d. using the proper kind of storm windows and doors.

soffit ✓ 24. Attic vents may be in the roof, ______________, or gable.

Canadian ✓ 25. The R-2000 home program is sponsored by the ______________ Home Builders Association.

The Arkansas Plan ✓ 26. The energy-saving home sponsored by the U.S. Department of Housing and Urban Development is known as the ______________ plan.

F ✓ 27. Homes in hot, humid climates should have windows that face east and west. (T or F)

T ✓ 28. Homes in hot, humid climates usually have a slab-on-grade foundation. (T or F)

F ✓ 29. Homes in hot, humid climates should use more framing members to reduce heat conduction. (T or F)

6 ✓

12 ✓ 30. The Arkansas plan house has ______________ inches of insulation in the walls and ______________ inches in the ceiling.

32.2% ✓ 31. In the Arkansas plan house, about ______________ percent of the energy saving was achieved by using the correct windows and doors.

active ✓ 32. The two kinds of solar energy systems for living units are passive and ______________.

Collector ✓ 33. In an active solar energy system, the major problems are posed by the ______________ and storage units.

Convection ✓ 34. In the passive system, heat circulates to the living space by conduction, ______________ and radiation.

5 p.49 ✓ 35. There are ______________ basic elements in a complete passive system.

isolated ✓ 36. There are three types of passive solar energy systems – direct gain, indirect gain, and ______________ gain.

Sunlight ✓ 37. With a photovoltaic panel, electricity can be produced directly from ______________.

Study Unit 4
Wood as a Building Material

_____Cambium_____ ✓

_____sapwood_____ ✓

_____Rings_____ ✓

_____T_____ ✓

_____deciduous_____ ✓

✗ _____F_____ ✓   Coniferous + deciduous

_____F_____ ✓

_____Fir -wrong_____ ✓

_____F_____ ✓

_____Flat_____ ✓

_____Kiln_____ ✓

_____T_____ ✓

_____F_____ ✓

_____19%_____ ✓

NC 12% _____18%_____ ✓

_____F_____ ✓

_____Saturation_____ ✓

_____F_____ ✓

_____T_____ ✓

_____meter_____ ✓

1. The substance located under the bark of trees that produces new wood is the _____.

2. The sap is carried to leaves by the _____.

3. A tree's age can be judged by counting the number of growth _____.

4. Softwoods, when classified by species, are sometimes harder than hardwoods. (T or F)

5. Another name for hardwoods is _____ trees.

6. Softwoods come from trees that shed their leaves. (T or F)

7. Earlywood is darker than latewood. (T or F)

8. Common hardwoods include (one wrong): a. fir; b. maple; c. cherry; d. oak.

9. Plain-sawed lumber does not tend to vary as much as quarter sawed. (T or F)

10. One common method for cutting boards is called plain sawed for hardwood and _____ grained for softwood.

11. The two common methods for drying lumber are _____ drying and air drying.

12. When seasoning rough lumber, stickers should be placed between the pieces. (T or F)

13. Oven drying of lumber takes longer than the natural method. (T or F)

14. After lumber is dried in air, the moisture content should be _____ percent or less.

15. Lumber dried in an oven should have a moisture content of less than _____ percent.

16. Lumber shrinks about the same in length as it does in width. (T or F)

17. When wood contains enough water to saturate the cell walls it is said to be at the fiber _____ point.

18. Once wood has been dried, its moisture content remains constant. (T or F)

19. Moisture content of wood can be determined by the oven drying method. (T or F)

20. Another way to check moisture content is with a moisture _____.

Building Code only applies to air dried wood 19% moisture content

(Continued on next page)

_____ *a* ✓

_____ *T*

_____ *lower* ✓

✓
3 a. _____ 3 _____

5 b. _____ 5 _____

9 c. _____ 9 _____

12 d. _____ 7 _____ *12*

6 e. _____ 6 _____

8 f. _____ 8 _____

13 g. _____ 2 _____

11 h. _____ 11 _____ *13*

7 i. _____ 13 _____ *7*

1 j. _____ 1 _____

4 k. _____ 4 _____

10 l. _____ 10 _____

15 m. _____ 15 _____

14 n. _____ 14 _____

2 o. _____ _____ *1*

_____ 3 ✓

_____ 300% ✓

_____ F ✓

_____ Stress ✓

✓
2 a. _____ 2 _____

3 b. _____ 3 _____

7 c. _____ 7 _____

6 d. _____ 6 _____

8 e. _____ 8 _____

1 f. _____ 1 _____

5 g. _____ 5 _____

4 h. _____ 4 _____

21. When a board is dried rapidly in an oven, the center will have *(one right)*: a. more moisture than the outside of the board; b. less moisture than the outside; c. the same amount of moisture; d. no moisture.

22. Pieces of higher and lower quality are found in the same grade of lumber. (T or F)

23. The best-quality lumber is found in the _____ part of the tree trunk, near the outside.

24. Match the items on the left with the descriptions on the right:

a. stain	1. any variation from true surface
b. knot	2. deviation flatwise from straight line
c. check	3. discoloration
d. pitch-pocket	4. dense, outside part of ring
e. decay	5. branch or limb embedded in tree
f. wane	6. caused by fungi
g. cup	7. grain separation between or through growth rings
h. crook	8. bark on edge or corner
i. shake	9. lengthwise grain separation due to seasoning
j. warp	10. separation of wood completely through to opposite surface
k. summerwood	11. edgewise deviation that goes full length
l. split	12. an opening containing pitch or bark
m. pitch	13. deviation flatwise across width
n. torn grain	14. wood torn out
o. bow	15. accumulation of resin

25. There are _____ grades of hardwood.

26. When a tree is first cut down, it can contain moisture ranging from 30 to _____ percent more than it does after drying.

27. All wood should be dried the same amount regardless of its use. (T or F)

28. Lumber labeled according to its strength and loadbearing quality is called _____ grade lumber.

29. Match the items at the left with the descriptions on the right:

a. tensile stress	1. applies to furniture in a home
b. compression	2. tends to make a piece longer
c. live load	3. squeezing or crushing
d. impact load	4. applied a large number of times
e. elasticity	5. forces that meet head on
f. static load	6. a sudden, sharp force
g. shear stress	7. weight applied to a house
h. fatigue load	8. stiffness of a piece and its resistance to bending

_F_ 30. Painting wood prevents it from taking on moisture. (T or F)

_T_ 31. If lumber is stored properly, it will not absorb much moisture. (T or F)

_F_ 32. Hardwood is sold in standard widths and lengths. (T or F)

_F_ 33. The nominal size of lumber and its actual size are the same. (T or F)

_12_ 34. A board that measures 2″ × 6″ × 12′ contains _____ board feet.

_8_ 35. A 2″ × 4″ that is 12′ long contains _____ board feet.

2x4x12 ÷ 12

millimeters (mm) 36. In the metric system, the unit for thickness and width of lumber is the _____.

meter (m) 37. The metric unit for length is the _____.

_2440_ 38. The metric measurement for a standard plywood sheet is 1220 mm by _____ mm.

Study Unit 5
Plywood and Composition Panels

_____ *veneer* ✓ 1. A thin sheet of wood that is sliced, sawed, or peeled from a log is called _____.

_____ *plies* ✓ 2. Plywood is made from layers and/or _____ of veneer or veneer and wood.

_____ *F* ✓ 3. The plywood layers are glued together with the grain running in the same direction. (T or F)

_____ *lumber* ✓ 4. There are two major types of hardwood plywood, namely, veneer core and _____ core.

_____ *T* ✓ 5. Logs or flitches must be softened or tenderized before cutting the veneer. (T or F)

_____ *T* ✓ 6. Most plywood used in the construction industry is made from softwood. (T or F)

_____ *particleboard* ✓ 7. The three kinds of plywood construction are veneer core, lumber core, and _____ core.

_____ *waterproof* ✓ 8. Exterior plywood must be put together with _____ glue.

_____ *T* ✓ 9. Exterior plywood construction requires the application of heat. (T or F)

_____ *T* 10. Interior plywood can be made without heat. (T or F)

_____ *structural* ✓ 11. Plywood that is made for special engineering use and meets such properties as tension and compression is _____ plywood.

_____ *T* ✓ 12. Plywood is always manufactured in an odd number of layers. (T or F)

_____ *D the lowest grade* ✓ 13. Plywood grades range from N to _____.

_____ *5* ✓ 14. The 70 different species of wood from which plywood is made are divided into _____ groups.

_____ *T* ✓ 15. Maple, birch, and Douglas fir are three of the strongest woods. (T or F)

_____ *T* ✓ 16. Choose plywood thickness according to its use. (T or F)

_____ *T* ✓ 17. The number of plies and the thickness of plywood are important factors in its use. (T or F)

_____ *T* ✓ 18. The front and back of a plywood piece can have different grades. (T or F) *CDX*

_____ *warp* ✓ 19. If plywood is stored on edge, it will _____.

_____ *T* ✓ 20. It is best to store plywood flat. (T or F)

_____ *F* ✓ 21. To saw plywood by hand, place the good face down. (T or F)

_____ *T* ✓ 22. Screws and nails hold well only in the face of plywood. (T or F)

(Continued on next page)

_____ T 23. The correct nail and screw size to use depends on the thickness of the plywood. (T or F)

_____ T 24. Composition panels are made from pieces of wood. (T or F)

_____ T 25. Materials for composition panels can be made from trees that are too small for lumber. (T or F)

_____ F 26. Urea resins are waterproof. (T or F)

_____ T 27. OSB is used primarily for siding and subflooring. (T or F)

_____ F 28. OSB will not shrink or swell with a change in humidity. (T or F)

_____ T 29. Waferboard is made of flakes of wood that are randomly aligned throughout the panel. (T or F)

_____ T 30. Hardboard is an all-wood panel manufactured from wood fibers. (T or F)

_____ C 31. Some characteristics of hardboard are (one wrong): a. exceptional strength; b. superior wear resistance; c. hard to work with ordinary tools; d. easy to paint.

_____ tempered 32. The three types of hardboard are standard, _____, and service.

_____ T 33. Hardboard classified as S1S has one surface smooth and the other rough. (T or F)

_____ b 34. Manufactured boards have the following advantages over basic building materials like wood (one wrong): a. they can be made to more sizes, shapes, and surfaces; b. they are heavier; c. they are high in insulating values; d. they do not tend to split, crack, or splinter.

_____ T 35. Hardboard, particleboard, plywood, insulating board, gypsum board, and plastic foam board are all examples of building boards. (T or F)

36. Match the hardboards at the left with the descriptions at the right:

2 a. _____ 2   a. embossed   1. most popular for indoor paneling
1 b. _____ 1   b. wood-grain   2. made to imitate a different
4 c. _____ 4   c. perforated   material
6 d. _____ 6   d. underlayment   3. for sound control
7 e. _____ 7   e. exterior siding   4. excellent for storage or display
3 f. _____ 3   f. acoustical   5. cut into patterns
5 g. _____ 5   g. filigree   6. a good subflooring
                                                             7. prefinished and primed at the factory

_____ T 37. Particleboard is made by combining wood flakes with adhesives. (T or F)

_____ F 38. Heat is not required in the manufacture of particleboard. (T or F)

_____ extruded 39. The two kinds of particleboard are mat-formed and _____.

_____ C 40. Particleboard has the following characteristics (one wrong): a. is a good base for plastic laminates; b. does not warp; c. takes a good paint finish; d. is stiff and strong.

_____ T 41. Particleboard is used in home construction to make kitchen counters, sink tops, and cabinets. (T or F)

_____ *underlayment* ✓42. The largest single use of particleboard in the construction industry is for floor _____.

_____ T ✓43. There is a wide variation in the types of particleboard manufactured. (T or F)

_____ T ✓44. Particleboard can be worked with standard wood tools. (T or F)

_____ F ✓45. Particleboard is always purchased prefinished. (T or F)

_____ T ✓46. Most ordinary wood joints can be made in particleboard. (T or F)

Study Unit 6
Framing Connectors and Engineered Wood

_____ b wrong _____ 1. Several important structural products include *(one wrong)*: a. laminated-veneer lumber; b. heavy timbers; c. glue-laminated beams; d. metal connectors.

_____ T _____ 2. Metal joist hangers can replace toenailed connections.

laminated veneer 3. A family of products made with wood veneer as the basic element is called _____ _____ lumber.

_____ LVL _____ 4. The letters used to identify laminated-veneer lumber are _____.

_____ 60' _____ 5. LVL products are available in lengths of _____ feet or more.

_____ T _____ 6. LVL products shrink and swell less than solid lumber.

F parallel-lamination 7. LVL products are cross-laminated.

I-joist 8. An I-beam is also called an _____.

_____ F _____ 9. I-beams can be used only in floor construction.

_____ T _____ 10. The webs of LVL I-beams are made from plywood or oriented-strand board.

waterproof adhesive 11. LVL I-beams use contact adhesive to attach webs and chords.

Radial arm saw 12. The easiest way of cutting an I-beam is with a _____ _____ saw.

_____ T _____ 13. I-beams used in floor construction can be nailed to the plate.

joist hangers 14. I-beams can be secured with metal _____ _____.

F 45° angle 15. Nails should be driven sideways (parallel to the laminations) into an I-beam chord.

F -doubled 16. I-beams can be tripled to form a header.

1½" 17. The web of a wood I-beam has _____ inch diameter pre-scored knockouts located approximately 12" on center.

plumbing + electrical 18. Knockouts in I-beams are used to create passages for _____ and _____ lines.

3/32" 19. Framing connectors of 12-gauge galvanized steel are approximately _____ inch (fractional) thick.

_____ F _____ 20. The chords of I-beams can be notched or drilled as needed.

laminated /a Rectangular 21. LVL headers are _____ in cross-section.

_____ F _____ 22. Holes can be cut in LVL headers.

waterproof 23. All laminated-veneer products use _____ adhesive.

(Continued on next page)

Answers (handwritten, left column):

24. glulam
25. T
26. F — dimension lumber
27. d — wrong
28. Camber
29. b — wrong
30. T
31. 3½ or 5½
32. T
33. T
34. F — joist
35. 16d
36. truss
37. T
38. F — too few
39. angular
40. F — nails

---

24. Glue-laminated beams are also called _____.

25. Glue-laminated beams are resistant to fire. (T or F)

26. Glue-laminated beams are made of layers of solid sawn lumber glued together. (T or F)

27. Glue-laminated beams can be used for *(one wrong)*: a. garage door; b. windows; c. patio door; d. only heavy construction.

28. A slight upward curve in the glulam, like the crown in a piece of lumber is called _____.

29. Three grades of glulam include *(one wrong)*: a. industrial; b. commercial; c. architectural; d. premium.

30. Premium-grade glulams are used where appearance is of prime importance. (T or F)

31. Glulams used for standard headers are either _____ inches or _____ inches wide.

32. A metal connector is a formed or stamped metal bracket. (T or F)

33. The builder must check local building codes to make sure certain metal connectors are approved for that application. (T or F)

34. A joint hanger is the most common metal connector. (T or F)

35. Use _____ nails to improve the strength of joist hangers.

36. Figure 6-1 shows a metal connector between the _____ and a double top plate.

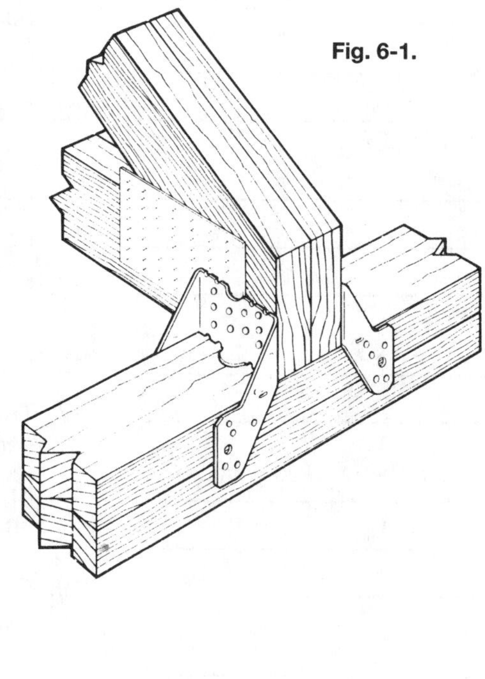

37. There are two methods of installing a joist hanger. (T or F)

38. The most common mistake when installing joist hangers is to use too many nails. (T or F)

39. Two common shapes of metal framing ties are the flat strap and the _____ shape.

40. Drywall screws are recommended to secure metal framing connectors. (T or F)

Fig. 6-1.

Name _____

Score: (47 possible) _____

Study Unit 7
Safety

_____ T _____

_____ T _____

_____ F _____

_____ F _____

_____ T _____

_____ Respirator _____

_____ T _____

_____ Kickback _____

_____ F _____

_____ T _____

_____ T _____

_____ C - wrong _____

_____ a - wrong _____

_____ d - wrong _____

_____ F _____

a. _____ Run _____

b. _____ interrupt _____

c. _____ power _____

d. _____ tools _____

e. _____ stop _____

f. _____ metal _____

1. Injuries are costly because they reduce the efficiency of the work force. (T or F)

2. Falling is the most common cause of injury on the job site. (T or F)

3. Older workers are more likely to be injured than younger workers. (T or F)

4. Follow this rule: "Lift with your back, not with your knees." (T or F)

5. Many building materials become slippery when wet or frosty. (T or F)

6. To protect yourself against chemical fumes, wear a _____.

7. Tilesetters should wear kneepads. (T or F)

8. When stock being cut on a table saw is thrown back at a high speed, this is called _____.

9. When using a portable circular saw, force the saw into the cut. (T or F)

10. Moisture can turn many materials into good conductors. (T or F)

11. GFCI is a fast-acting circuit breaker. (T or F)

12. Proper eye protection requires wearing *(one wrong)*: a. a mask; b. safety goggles; c. sunglasses; d. a safety shield.

13. The correct dress for building construction includes *(one wrong)*: a. jewelry; b. sleeves of shirts or jackets buttoned; c. short hair or a hair protector; d. working attire.

14. Complete safety attire for building construction includes the following *(one wrong)*: a. hard hat; b. safety shoes; c. safety eye protector; d. tie.

15. When lifting heavy objects, use the muscles in your back. (T or F)

16. Follow these general safety practices:

    a. Always walk; do not _____.

    b. Never talk to or _____ anyone who is working on a machine.

    c. Remove the power plug or turn off the _____ supply to a machine when changing cutters or blades.

    d. Never leave _____ or pieces of stock lying on the table surface of a machine before using them.

    e. When finished with a machine, turn off power and wait until the cutter has come to a complete _____ before leaving.

    f. Always carefully check stock for knots, splits, _____ objects and other defects before machining.

(Continued on next page)

g. _____understand_____

h. _____shards_____

i. _____fingers_____

j. _____clear_____

k. _____off_____

l. _____brush_____

m. _____cutting_____

n. _____sharp_____

o. _____noises_____

a. _____slip_____

b. _____scaffolding_____

c. _____ladders_____

_____OSHA_____

_____1971_____

_____F-daily_____

_____T_____

_____T or Remove_____

_____T_____

_____b_____

_____14_____

g. Do not use a machine until you _____ it thoroughly.

h. Use _____ on power equipment.

i. Always keep your _____ away from the moving cutting edges.

j. Keep the floor around the machine _____.

k. Make all adjustments with the power _____.

l. Always use a _____ to clean the table surface.

m. Keep your eyes focused on where the _____ action is taking place.

n. Make sure that tools are _____.

o. Report strange _____ or faulty operation of machines to the instructor.

17. Complete the following about avoiding falls:

   a. Learn to watch your _____; avoid objects that could trip you.

   b. Check _____ and temporary walkways before walking on them.

   c. Use only _____ which are in good condition and set up properly.

18. The correct abbreviation for the Federal Occupational Safety and Health Act is _____.

19. The Federal Occupational Safety and Health Act was passed by the Congress of the United States in April of _____.

20. Scraps and rubbish at the job site should be removed weekly. (T or F)

21. Materials and equipment should be kept in neat, straight stacks. (T or F)

22. A board with protruding nails should have the nails bend downward. (T or F)

23. Aisles and walkways should be clear of tools, materials, and debris. (T or F)

24. Long pieces of material should be carried by (one right): a. one person; b. two people; c. three people; d. several people.

25. The recommended extension cord size for use with portable power tools when the tool has an amperage rating of 8 and the cord length is 100 feet is _____ A. W. G.

a. _____ water _____

b. _____ grounded _____

c. _____ eye _____

d. _____ plug _____

e. _____ switch _____

_____ Guard _____

26. Complete the following concerning safety rules for portable power tools:

a. Never use portable power tools in contact with _____.

b. Portable power tools should be properly _____.

c. Always wear approved _____ protection.

d. Always disconnect the power _____ when the work is completed.

e. Be sure that the _____ is in the off position before connecting to a power plug.

27. Whenever the drawing shown here in Fig. 7-1 appears with an illustration, a _____ must be used for the operation shown.

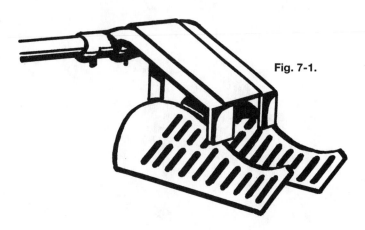

Fig. 7-1.

(Continued on next page)

To be used with Unit 10 | Name ____________________

Pages 109-121 | Score: (169 possible) ______

Study Unit 8

Hand Tools

LAYOUT, MEASURING, AND CHECKING DEVICES

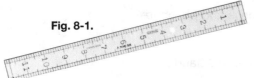

Fig. 8-1.

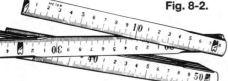

Fig. 8-2.

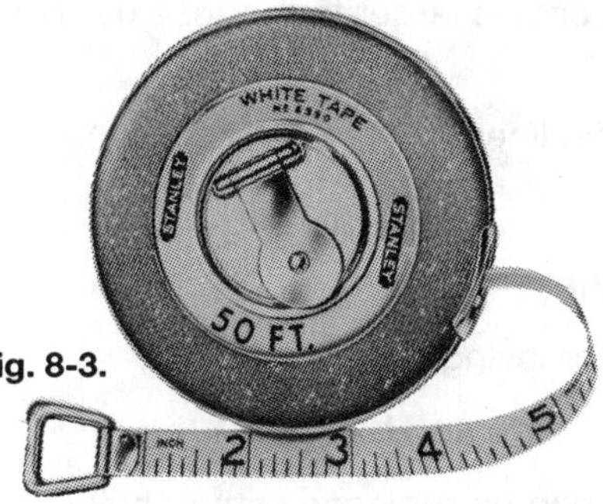

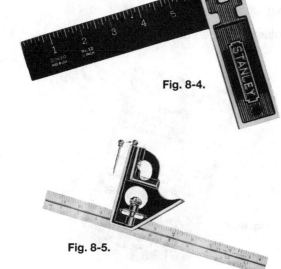

Fig. 8-3.

Fig. 8-4.

Fig. 8-5.

1. The hand tools shown here are numbered 8-1 through 8-6. Place these numbers in the blanks provided, matching tools with their correct names, descriptions, and uses.

_________ Zig-zag rule

_________ Bench rule

_________ Steel tape has a hook on the end that adjusts to true zero

_________ Used to adjust dividers

_________ Try square

_________ A folding rule

_________ A good tool for measuring an angle

_________ Sliding T bevel

_________ This should never be used as a straightedge

_________ Combination square

_________ Used to measure or transfer an angle between 0 and 180 degrees

_________ Flexible tape rule

_________ The blade on this tool slides along the handle

_________ Comes in lengths from 6′ to 100′ (2 m to 50 m)

_________ Used to mark and test a 45-degree miter

_________ Measures irregular and regular shapes

_________ Used to check adjacent surfaces for squareness

_________ Measures distances, using tool flat or on edge

_________ Makes simple measurements

_________ Makes accurate inside measurements

_________ Makes lines across face or edge of stock

Fig. 8-6.

(Continued on next page)

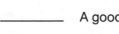

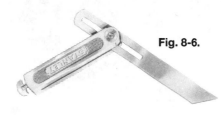

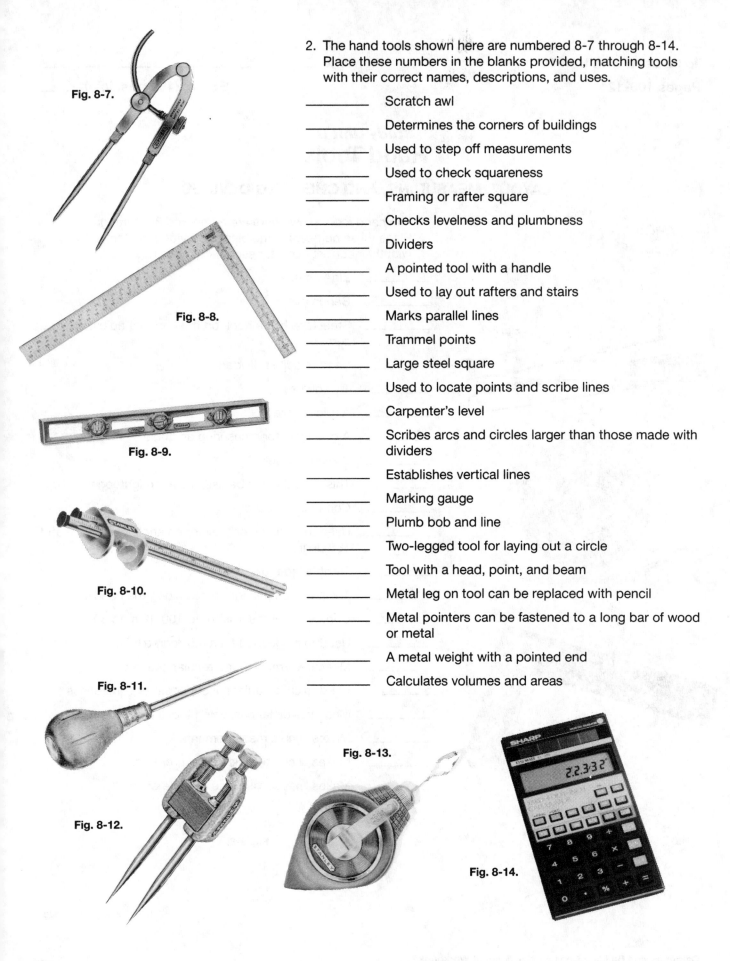

Fig. 8-7.

Fig. 8-8.

Fig. 8-9.

Fig. 8-10.

Fig. 8-11.

Fig. 8-12.

Fig. 8-13.

Fig. 8-14.

2. The hand tools shown here are numbered 8-7 through 8-14. Place these numbers in the blanks provided, matching tools with their correct names, descriptions, and uses.

_____ Scratch awl

_____ Determines the corners of buildings

_____ Used to step off measurements

_____ Used to check squareness

_____ Framing or rafter square

_____ Checks levelness and plumbness

_____ Dividers

_____ A pointed tool with a handle

_____ Used to lay out rafters and stairs

_____ Marks parallel lines

_____ Trammel points

_____ Large steel square

_____ Used to locate points and scribe lines

_____ Carpenter's level

_____ Scribes arcs and circles larger than those made with dividers

_____ Establishes vertical lines

_____ Marking gauge

_____ Plumb bob and line

_____ Two-legged tool for laying out a circle

_____ Tool with a head, point, and beam

_____ Metal leg on tool can be replaced with pencil

_____ Metal pointers can be fastened to a long bar of wood or metal

_____ A metal weight with a pointed end

_____ Calculates volumes and areas

SAWING TOOLS

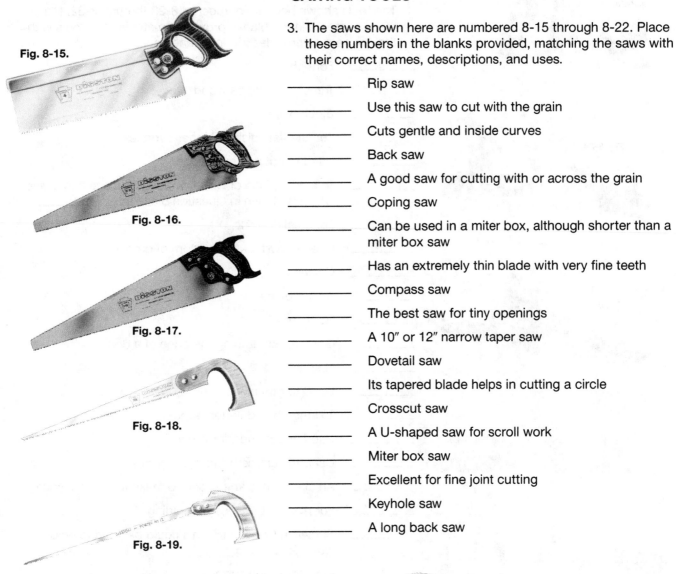

Fig. 8-15.

Fig. 8-16.

Fig. 8-17.

Fig. 8-18.

Fig. 8-19.

3. The saws shown here are numbered 8-15 through 8-22. Place these numbers in the blanks provided, matching the saws with their correct names, descriptions, and uses.

_____ Rip saw

_____ Use this saw to cut with the grain

_____ Cuts gentle and inside curves

_____ Back saw

_____ A good saw for cutting with or across the grain

_____ Coping saw

_____ Can be used in a miter box, although shorter than a miter box saw

_____ Has an extremely thin blade with very fine teeth

_____ Compass saw

_____ The best saw for tiny openings

_____ A 10″ or 12″ narrow taper saw

_____ Dovetail saw

_____ Its tapered blade helps in cutting a circle

_____ Crosscut saw

_____ A U-shaped saw for scroll work

_____ Miter box saw

_____ Excellent for fine joint cutting

_____ Keyhole saw

_____ A long back saw

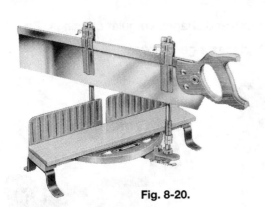

Fig. 8-20.

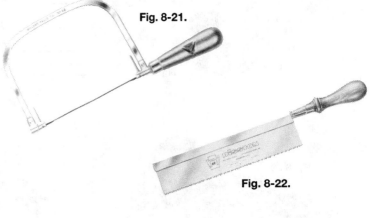

Fig. 8-21.

Fig. 8-22.

(Continued on next page)

EDGE-CUTTING TOOLS

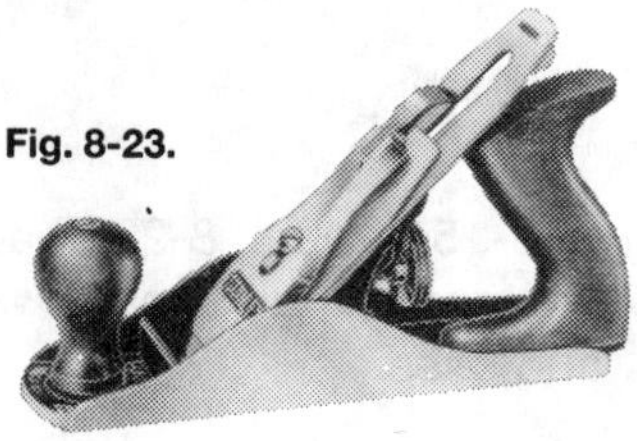

Fig. 8-23.

Fig. 8-24.

Fig. 8-25.

Fig. 8-26.

Fig. 8-27.

4. The tools shown here are numbered 8-23 through 8-32. Place these numbers in the blanks provided, matching the tools with their correct names, descriptions, and uses.

_____ Fore plane

_____ Trims and shapes wood

_____ Block plane

_____ For fine flat finish or longer surfaces

_____ Used to cut and trim wood

_____ To hold a piece of wood, you would place it in a vise, which is shown in Question 5 as Fig. _____

_____ For smaller work

_____ Used to surface the bottom of grooves

_____ Chisel

_____ An 18″ plane

_____ Router plane

_____ Good for smoothing the edge of a door

_____ Hatchet

_____ Used to trim pieces

_____ To smooth and flatten edges

_____ Used to cut and trim veneer

_____ Used to surface bottom of dadoes

_____ An all-purpose knife used to make accurate layouts

_____ For planing the ends of molding

_____ Shown in Question 5, this could be used to hold wood for planing.

_____ Ideal for rough surfaces

_____ Jack plane

_____ Cuts grooves and shapes irregular openings

_____ A 7″ to 9″ plane for general use

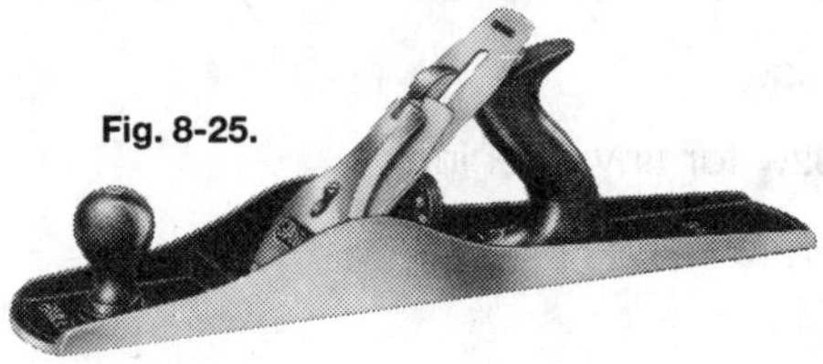

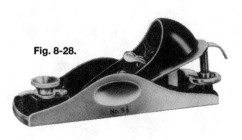

Fig. 8-28.

Fig. 8-29.

Fig. 8-30.

Fig. 8-31.

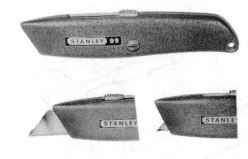

Fig. 8-32.

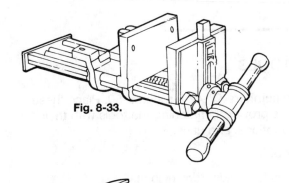

Fig. 8-33.

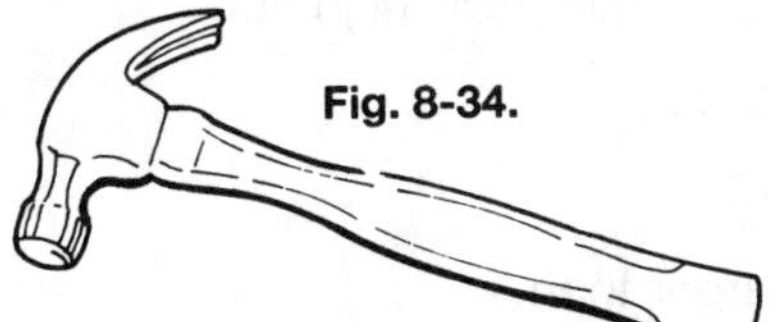

5. The tools shown here are numbered 8-33 through 8-41. Place these numbers in the blanks provided, matching the tools with their correct names, descriptions, and uses.

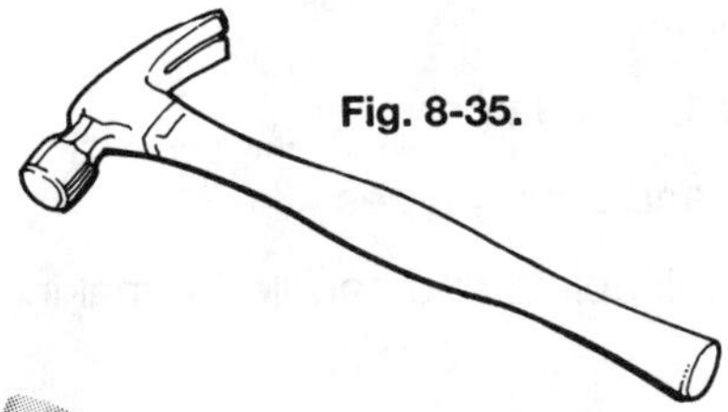

_____ For driving nails below wood surfaces

_____ Screwdriver

_____ Has claw designed primarily for removing nails

_____ Vise

_____ Striking tool used when steel face would damage the surface of the material

_____ Comes with slotted or Phillips head

_____ Pneumatic nailer-stapler

_____ Ripping bar

_____ Hammer designed to pry nailed pieces apart

_____ May attach permanently to workbench

_____ Nail set

_____ Mallet

_____ Drives staples with spring-driven plunger

_____ Claw hammer

_____ Rip hammer

_____ For attaching insulation

_____ Available in lengths up to 8′

_____ Holds work for planing

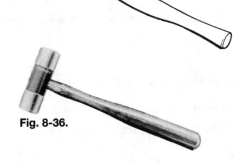

Fig. 8-34.

Fig. 8-35.

Fig. 8-36.

Fig. 8-37.

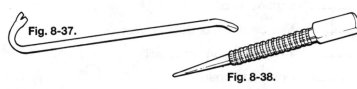

Fig. 8-38.

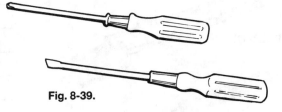

Fig. 8-39.

(Continued on next page)

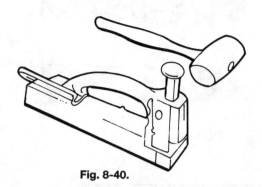

Fig. 8-40.

Fig. 8-41.

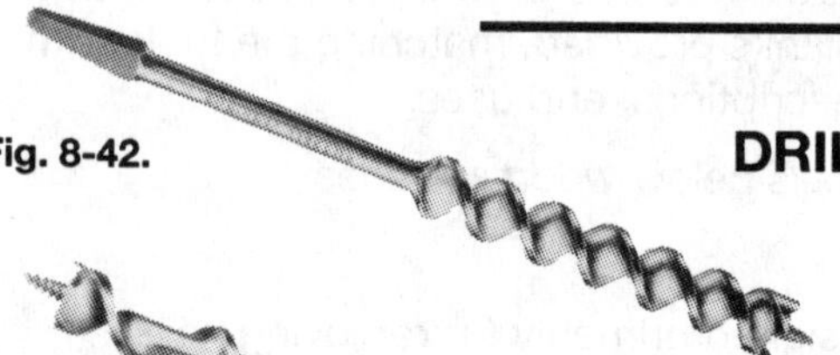

Fig. 8-42.

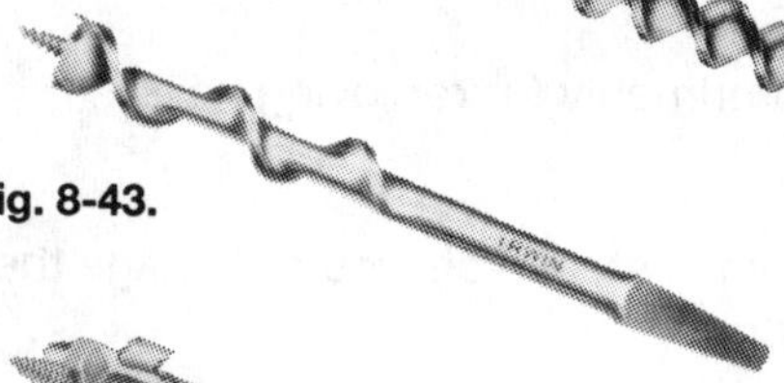

Fig. 8-43.

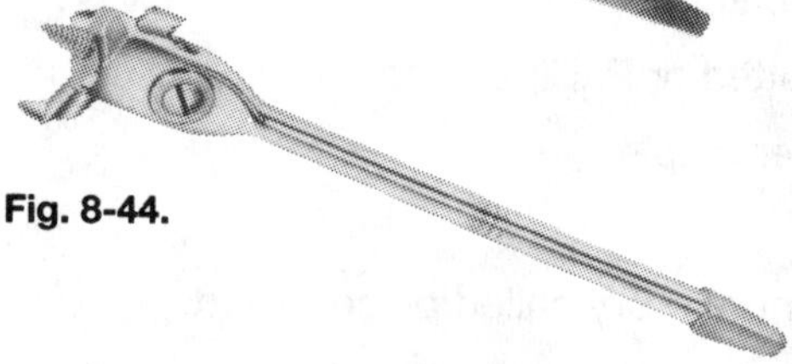

Fig. 8-44.

Fig. 8-45.

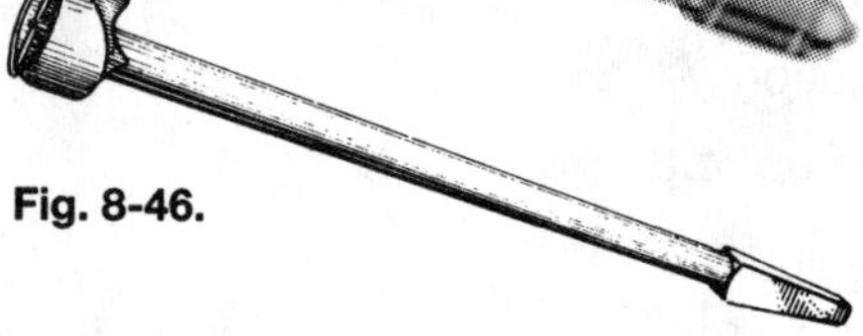

Fig. 8-46.

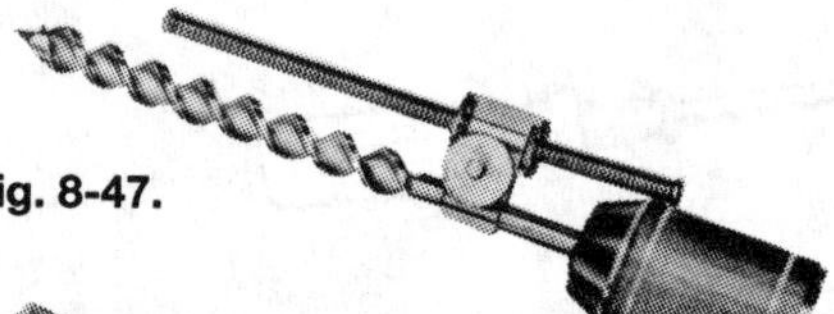

Fig. 8-47.

Fig. 8-48.

DRILLING AND BORING TOOLS

6. The tools shown are numbered 8-42 through 8-50. Place these numbers in the blanks provided, matching the tools with their correct names, descriptions, and uses.

_____ Foerstner bit

_____ Bores holes for making dowel joints

_____ Has a 3-jaw chuck

_____ Auger bit

_____ Bores holes larger than 1″

_____ Plain and ratchet types

_____ Hand drill

_____ Bores holes 1/4″ or larger

_____ Solid clamp and spring types

_____ A tool with drill points and handle for making many small holes

_____ Brace

_____ Will bore a shallow hole with a flat bottom

_____ Bit or depth gauge

_____ Holds twist-drills for drilling small holes

_____ Automatic drill

_____ Expansion bit

_____ May be single or double twist

_____ Twist drill or bit stock drill

_____ Holds and operates bits

_____ Dowel bit

Fig. 8-49.

_____ Holds cutters of different sizes

_____ Will bore holes in thin stock and end grain

Fig. 8-50.

METALWORKING TOOLS

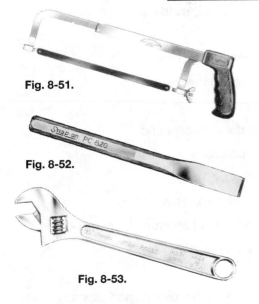

Fig. 8-51.

Fig. 8-52.

Fig. 8-53.

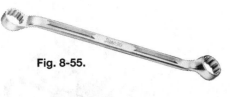

Fig. 8-54.

Fig. 8-55.

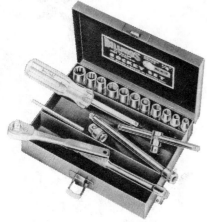

Fig. 8-56.

7. The tools here are numbered 8-51 through 8-56. Place these numbers in the blanks provided, matching the tools with their correct names, descriptions, and uses.

_____ Used to assemble and disassemble machinery

_____ Adjustable wrench

_____ Uses replaceable metal-cutting blades

_____ Cuts off a rivet or nail

_____ Hacksaw

_____ Makes adjustments on machines where there is plenty of clearance

_____ A metal wrench with two enclosed ends

_____ Socket wrench set

_____ Used to install and replace knives and blades

_____ Open-end wrench

_____ Cuts all types of metal fasteners, hardware, and metal parts

_____ Makes adjustments where there is limited space for movement

_____ Fits many sizes of bolts and nuts

_____ Cold chisel

_____ Has U-shaped frame

_____ Has a specially hardened and tempered cutting edge

_____ A nonadjustable wrench

_____ A series of sockets using a variety of handles

_____ Box wrench

_____ Will get a tight or rusted nut started

_____ For variety of work, a complete set is needed

(Continued on next page)

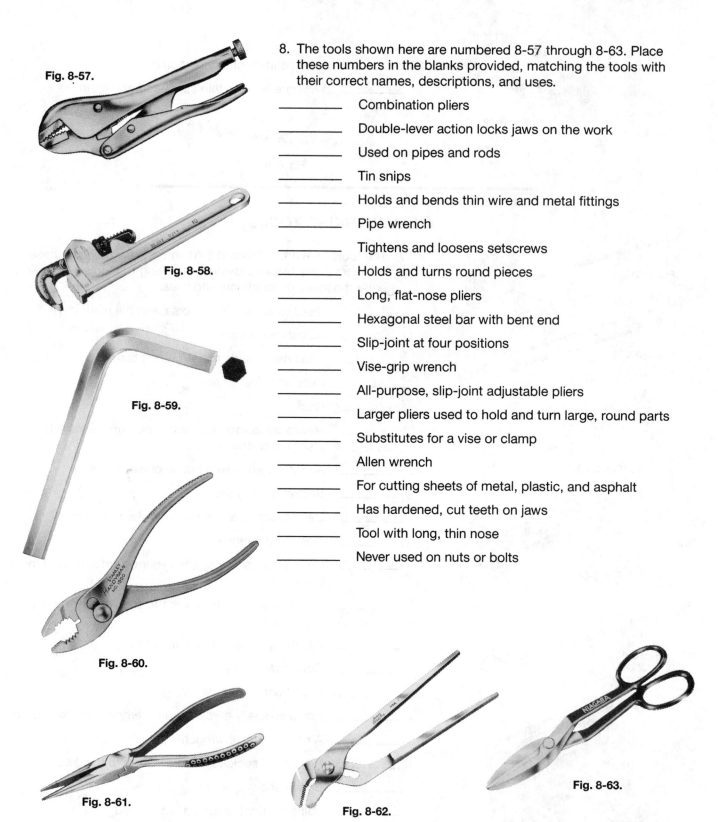

Fig. 8-57.

Fig. 8-58.

Fig. 8-59.

Fig. 8-60.

Fig. 8-61.

Fig. 8-62.

Fig. 8-63.

8. The tools shown here are numbered 8-57 through 8-63. Place these numbers in the blanks provided, matching the tools with their correct names, descriptions, and uses.

_____ Combination pliers

_____ Double-lever action locks jaws on the work

_____ Used on pipes and rods

_____ Tin snips

_____ Holds and bends thin wire and metal fittings

_____ Pipe wrench

_____ Tightens and loosens setscrews

_____ Holds and turns round pieces

_____ Long, flat-nose pliers

_____ Hexagonal steel bar with bent end

_____ Slip-joint at four positions

_____ Vise-grip wrench

_____ All-purpose, slip-joint adjustable pliers

_____ Larger pliers used to hold and turn large, round parts

_____ Substitutes for a vise or clamp

_____ Allen wrench

_____ For cutting sheets of metal, plastic, and asphalt

_____ Has hardened, cut teeth on jaws

_____ Tool with long, thin nose

_____ Never used on nuts or bolts

Name _____

Score: (32 possible) _____

Study Unit 9
Portable Circular Saw

1. Name the parts of the portable circular saw shown in Fig. 9-1:

a. _____

b. _____

c. _____

d. _____

e. _____

f. _____

g. _____

h. _____

Fig. 9-1.

2. When using the portable circular saw:

a. _____

b. _____

c. _____

d. _____

e. _____

f. _____

g. _____

_____

_____

_____

   a. Make sure the blade _____ are sharp and set correctly.

   b. Always keep the _____ in place.

   c. Disconnect the _____ source to change a blade.

   d. Allow the saw to reach _____ speed before starting a cut.

   e. Keep your _____ clear of the cutting line.

   f. Never stand in _____ with the cut.

   g. Wait until the blade has _____ before setting the saw down.

3. In cutting plywood with a portable circular saw, the good side should be _____.

4. A 10" circular saw would have a blade with a diameter of _____ inches.

5. The size of a portable circular saw is determined primarily by the _____ diameter, although the size of the motor is also important.

(Continued on next page)

6. A protractor attachment for a portable circular saw is ideal for *(one wrong)*: a. angle cutting; b. miter cutting; c. compound angle cutting; d. circle cutting.

_____

7. The cutting action of the portable circular saw is similar to that of the circular saw. (T or F)

_____

8. The saw should be allowed to reach half speed before starting a cut. (T or F)

_____

9. The saw blade should clear the bottom of the work by about _____ inch.

_____

10. An example of a pocket cut is the opening in a counter top for a sink. (T or F)

_____

11. It is possible to cut a miter freehand with a portable circular saw. (T or F)

_____

12. Ripping with a portable circular saw can be done freehand. (T or F)

_____

13. To make a _____ cut with a portable circular saw, the shoe is tilted to the desired angle after loosening the wing nut or handle.

_____

14. The portable cutoff table for use with a portable circular saw can be used to cut flat miters but not edge miters. (T or F)

_____

15. _____ miter stops are used as a guide when cutting flat miters on a cutoff table.

_____

16. _____ miters are cut on the cutoff table by tilting the portable circular saw to the desired angle.

_____

17. A portable circular saw can be equipped with a dustbag. (T or F)

_____

18. Two types of circular saws are the _____ saw and the worm-drive saw.

_____

19. A worm-drive saw has high blade speed and lower torque. (T or F)

Study Unit 10
Table Saw

_____

_____

_____

_____

_____

_____

_____

_____

1. Make all adjustments on the saw *(one right)*: a. when the power is off and the blade is motionless; b. while the power is on; c. during the cutting operation; d. just after the switch is turned off.

2. Scraps should be removed from around the saw blade by *(one wrong)*: a. pushing the scraps away with a push stick; b. lifting the guard with one hand and removing the scraps with the other while the machine is running; c. removing the scraps after the machine has come to a dead stop; d. using a brush to remove the scraps.

3. If the cut you are making does not permit the use of the regular guard, use the feather board or a _____ guard.

4. Always use a blade that is _____.

5. Do not saw _____ material on the table saw.

6. The size of the table saw is determined by *(one right)*: a. height of the saw; b. dimensions of the table; c. diameter of the recommended blade; d. size of the arbor.

7. The table of a variety saw is permanently fastened in the horizontal position. (T or F)

8. In selecting a saw blade, check the following *(one wrong)*: a. diameter of the blade; b. correct size of hole for arbor; c. kind of blade; d. brand name only.

a. _____

b. _____

c. _____

d. _____

e. _____

f. _____

9. Match the blades at the left with the descriptions on the right:

a. ripsaw	1. used primarily for trimming stock to length and squaring
b. hollow ground	
c. plywood saw	2. has chisel-like teeth
d. cutoff	3. used for a great variety of cutting
e. easy-cut	4. made of specially tempered steel
f. combination saw	5. practically eliminates kickback, so considered to be the safest blade
	6. used for fine cabinetwork

_____

_____

_____

10. A combination blade can be used for both ripping and crosscutting. (T or F)

11. An accessory that will cut all widths of grooves is called a *(one right)*: a. dado head; b. molding head; c. cutter head; d. grooving head.

12. The saw blade should protrude above the stock about *(one right)*: a. 1/4"; b. 1/8"; c. 1/2"; d. 1".

(Continued on next page)

_____

13. When removing a saw blade, remember the following about arbor threads *(one right)*: a. they may be either right-hand or left-hand threads, depending on the commercial make of the machine; b. they are left-hand threads, and you must turn the nut counterclockwise to loosen it; c. they are right-hand threads, and you must turn the nut clockwise to loosen it; d. they are double threads.

a. _____

14. Identify the parts of the saw shown in Fig. 10-1:

b. _____

c. _____

d. _____

e. _____

f. _____

g. _____

h. _____

i. _____

j. _____

k. _____

l. _____

m. _____

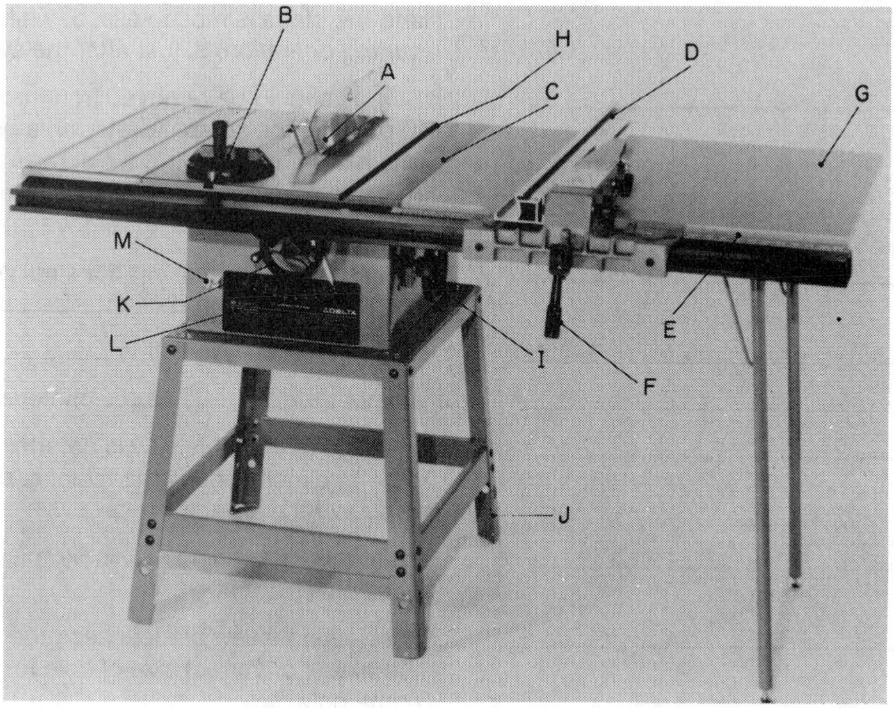

Fig. 10-1.

_____

15. For all ripping operations, use a ripping _____ to guide the work.

_____

16. When ripping a wide board, apply pressure against the fence with the right hand and push the board forward with the left. (T or F)

_____

17. For ripping a board longer than 6′ to 8′, you should have a helper or a _____ stand.

_____

18. When ripping narrower stock, the _____ stick is used to take the place of your right hand.

_____

19. To rip extremely thick wood, cut partway through from opposite sides. (T or F)

_____

20. For crosscutting operations, the _____ gauge is used.

_____

21. To prevent the stock from moving while the cut is made, use a _____ rod.

_____

22. Plywood presents special cutting problems because of its construction. (T or F)

_____ 23. When cutting plywood, always place the stock on the table with the _____ side up.

_____ 24. When cutting plywood, use one of the following techniques (one wrong): a. reverse the miter gauge to start the cut; b. clamp a straightedge to the plywood; c. cut the plywood freehand; d. use the ripping fence as a guide.

_____ 25. When using a ripping fence to cut identical pieces to length, always use a _____ block.

_____ 26. To cut a chamfer or bevel with the grain, the blade is tilted and the work held against the _____.

a. _____ 27. Name the cuts and joint cuts shown in Fig. 10-2.

b. _____

c. _____

d. _____

e. _____

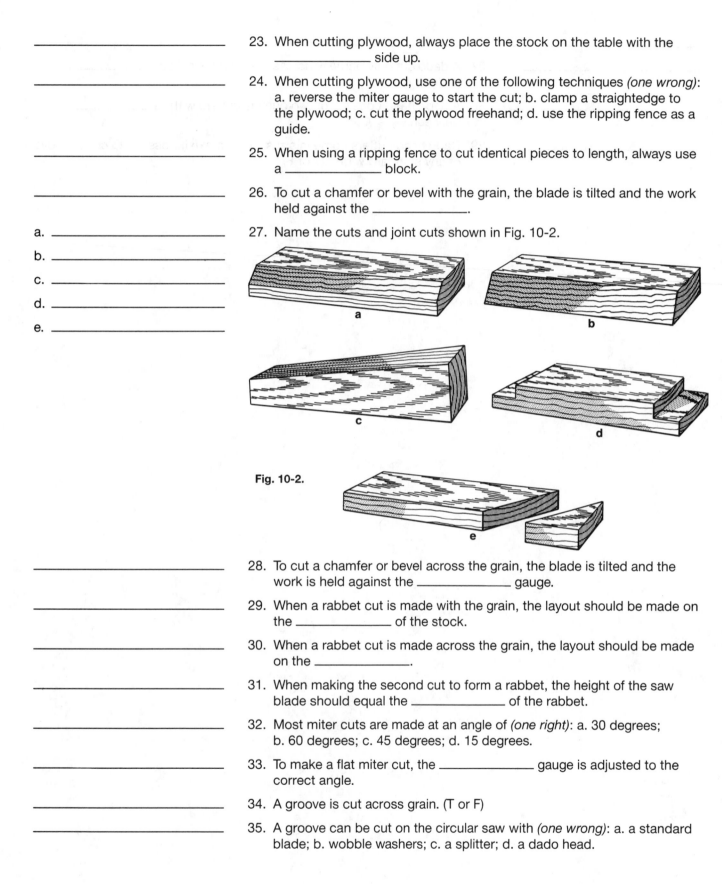

Fig. 10-2.

_____ 28. To cut a chamfer or bevel across the grain, the blade is tilted and the work is held against the _____ gauge.

_____ 29. When a rabbet cut is made with the grain, the layout should be made on the _____ of the stock.

_____ 30. When a rabbet cut is made across the grain, the layout should be made on the _____.

_____ 31. When making the second cut to form a rabbet, the height of the saw blade should equal the _____ of the rabbet.

_____ 32. Most miter cuts are made at an angle of (one right): a. 30 degrees; b. 60 degrees; c. 45 degrees; d. 15 degrees.

_____ 33. To make a flat miter cut, the _____ gauge is adjusted to the correct angle.

_____ 34. A groove is cut across grain. (T or F)

_____ 35. A groove can be cut on the circular saw with (one wrong): a. a standard blade; b. wobble washers; c. a splitter; d. a dado head.

(Continued on next page)

_____   36. A dado is cut with the grain. (T or F)

_____   37. A dado cut only partway across the board is called a _____ dado or gain.

_____   38. A dado head consists of two outside cutters with _____ in between.

_____   39. Dado head cutters come in sizes which make it possible to cut a groove of any standard width. (T or F)

Name _____

Score: (41 possible) _____

Study Unit 11
Radial-Arm Saw

_____

_____

_____

_____

1. The radial-arm saw can be used for ripping, dadoing, grooving, and various combinations of these cuts. (T or F)

2. One advantage of the radial-arm saw is that a long board can be easily cut into shorter lengths by sliding the board across the table after cutting. (T or F)

3. When installing a saw blade, do the following (*one wrong*): a. remove the wing nut to remove the guard; b. raise the blade so it will clear the table top; c. hold the blade with a block of wood; d. be sure the teeth at the bottom point away from you and toward the column.

4. The teeth of the saw blade should point in the direction of rotation. (T or F)

5. Name the parts of the radial-arm saw as shown in Fig. 11-1.

a. _____

b. _____

c. _____

d. _____

e. _____

f. _____

g. _____

h. _____

i. _____

j. _____

k. _____

l. _____

m. _____

n. _____

o. _____

p. _____

q. _____

r. _____

_____

Fig. 11-1.

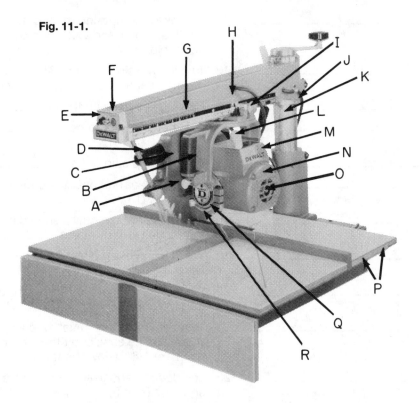

6. In crosscutting, the teeth of the saw should be about the following distance below the surface of the table (*one right*): a. 1/32"; b. 1/16"; c. 1/8"; d. 1/4".

(Continued on next page)

7. Crosscutting is done with the radial arm at _____ angles to the guide fence.

8. Figure 11-2 shows the basic ways to _____ on the radial-arm saw.

SAW FEED

Fig. 11-2.

GUIDE FENCE

STOCK

TABLE

THRUST

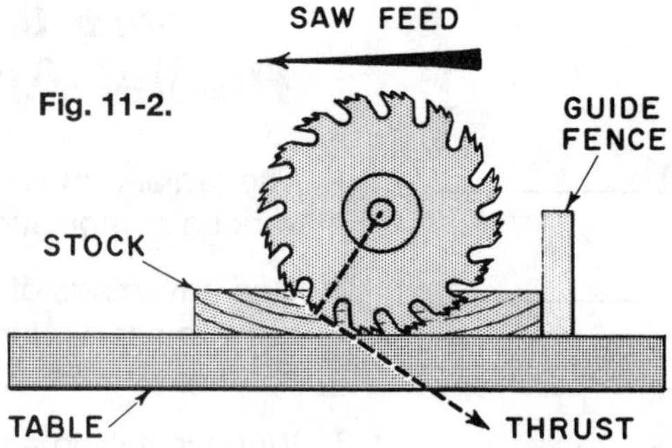

9. In crosscutting, the anti-kickback device should be about _____ inch above the work surface.

10. For crosscutting, as well as mitering, dadoing, and similar operations, the work is held firmly to the table and against the _____ fence.

11. Bevel cuts are made by tilting the _____ to the right or left to the correct angle.

12. To make a compound miter cut, the following two must be adjusted (one right): a. motor and yoke; b. motor and arm; c. arm and yoke; d. arm and table.

13. To cut a six-sided miter with the work tilted at 15 degrees, the blade must be set at _____ degrees.

14. Ripping is done with the blade _____ to the guide fence.

15. The motor is said to be _____ when it is away from the column.

16. The motor is said to be _____ when it is toward the column.

17. For ripping operations, the cutting head is stationary and the workpiece is moved. (T or F)

18. When ripping, the guard should be adjusted so that the infeed end clears the work by about _____ inch.

19. When using the radial-arm saw for ripping, the stock can be fed from either end. (T or F)

20. When ripping, make sure that the blade is rotating downward and away from you. (T or F)

21. To rip an angle, you must (one wrong): a. position the motor for ripping; b. elevate the saw; c. tilt the motor in the yoke; d. lower the saw until the teeth are 1/8″ below the table.

22. The dado heads used on a circular saw can also be used on the radial-arm saw. (T or F)

23. The saw teeth on a dado head should point the same way as with a saw blade. (T or F)

24. On most radial-arm saws (one wrong): a. one revolution of the elevation crank lowers the blade 1/8″; b. three revolutions lower the blade 1/16″; c. two revolutions lower the blade 1/4″; d. four revolutions lower the blade 1/2″.

Study Unit 12
Power Miter Saw

1. Follow these safety rules:

   a. _____

   b. _____

   c. _____

   d. _____

   e. _____

   f. _____

   g. _____

   h. _____

   _____

   a. Follow _____ instructions.

   b. Do not disable the blade _____.

   c. Make adjustments only after the _____ has stopped moving.

   d. Before changing blades, _____ the miter saw.

   e. Wear _____ glasses.

   f. Wear _____ protection as the saw has a high-pitched whine.

   g. Make sure the stock is _____ supported by the saw table.

   h. The saw should be equipped with a blade-_____.

2. The power miter saw is sometimes called a power miter _____.

   a. _____

   b. _____

   c. _____

   d. _____

   e. _____

   f. _____

3. Identify the parts of the power miter box shown in Fig. 12-1.

Fig. 12-1.

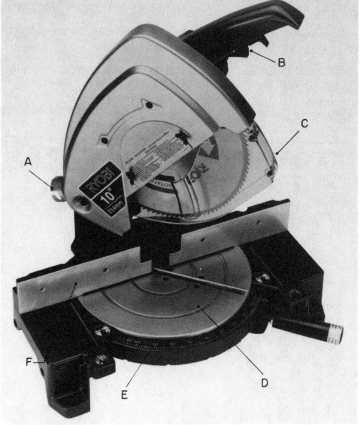

(Continued on next page)

4. The machine can cut miters either right or left. (T or F)

5. The power miter saw can be used for ripping. (T or F)

6. Saw blades range in size from 8" to _____".

7. The saw can be used to cut _____ extrusions.

8. A 2" × 4" can be cut at a 45-degree angle. (T or F)

9. When cutting flat pieces that are bowed, place the concave edge against the fence. (T or F)

10. With the correct kind of blade, the power miter saw can be used to cut the following (one wrong): a. plastic pipe; b. aluminum extrusions; c. stainless steel; d. plywood.

11. When cutting aluminum, a _____ wax should be applied to the side of the blade.

12. The material positioned against the saw fence as in Fig. 12-2 is correct. (T or F)

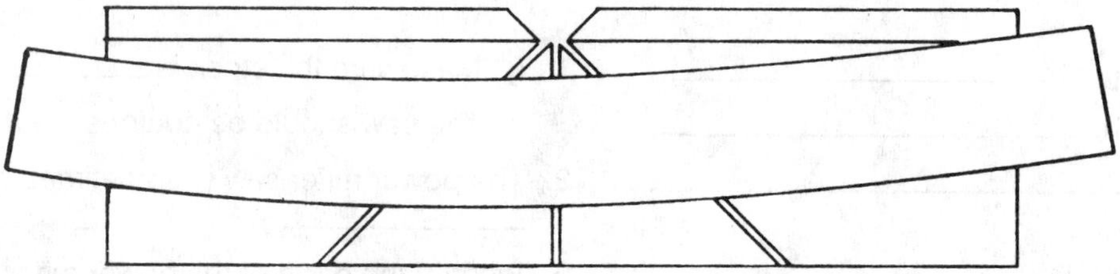

Fig. 12-2.

13. The saw has positive stops at 45 degrees and 90 degrees. (T or F)

14. A filler block can be used to hold the _____ molding in the correct position.

15. The blade of a compound-miter saw is mounted so that the blade can be _____.

16. A compound-miter saw can make a miter cut and a bevel cut at the same time. (T or F)

17. When mitering a 45-degree molding, the saw should be set at a 30-degree bevel angle and a _____-degree angle.

Study Unit 13
Portable Electric Drill

_____

_____

_____

_____

1. Portable electric drills equipped with a variable speed motor can be used as a screwdriver by inserting a screwdriver bit. (T or F)

2. The most common sizes of electric hand drills are the 1/4" and the _____ inch.

3. When using a 1/4", portable electric drill to hold a twist drill when drilling plastics, a high speed is used. (T or F)

4. To insert wood screws, the combination drill and countersink can be used. (T or F)

5. Name the parts of a 1/4" portable drill as shown in Fig. 13-1.

a. _____

b. _____

c. _____

d. _____

e. _____

f. _____

g. _____

h. _____

i. _____

j. _____

k. _____

l. _____

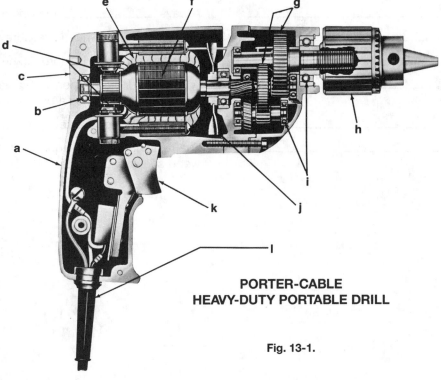

PORTER-CABLE
HEAVY-DUTY PORTABLE DRILL

Fig. 13-1.

6. When driving screws with a portable electric drill, use a _____ speed.

7. The combination drill will drill the pilot hole, shank hole, and _____ in one operation.

8. The maximum head diameter of a number 7 screw is _____ inches.

_____

_____

_____

(Continued on next page)

_____ 9. The fractional-size drill that should be used for drilling the shank hole for a number-10 wood screw is _____ inch.

_____ 10. Plastic drill housings are called double-insulated housings. (T or F)

_____ 11. A cordless electric drill is powered by a _____ battery.

_____ 12. A portable tool used to drill holes in masonry materials is called a _____ drill.

_____ 13. Screw guns can be used to drill holes. (T or F)

_____ 14. Most twist drills are made of high-speed steel. (T or F)

_____ 15. A bit that has a large spur point and horizontal cutting surfaces is called a _____ bit.

_____ 16. Brad point bits have a center _____ that prevents the bit from "walking" as the hole is started.

_____ 17. Bits designed to cut deep holes quickly through wood are called _____ bits.

_____ 18. There are special bits for boring holes in glass, ceramic tile, and mirrors. (T or F)

_____ 19. Always _____ the drill or bit when changing cutting tools.

_____ 20. When drilling, too much pressure will make the bit dull. (T or F)

_____ 21. A Phillips screw bit is more difficult to use than a straight bit. (T or F)

_____ 22. A portable drill equipped with a variable speed control and reversing switch can be used as a powerful _____.

_____ 23. A number-24 wood screw has _____ threads per inch.

_____ 24. The nearest fractional equivalent of the root diameter of a number-12 wood screw is _____ inch.

Study Unit 14
Pneumatic Nailers and Staplers

1. Complete the following safety precautions:

a. _____

_____

b. _____

c. _____

d. _____

e. _____

f. _____

g. _____

h. _____

i. _____

j. _____

k. _____

_____

_____

_____

_____

_____

a. Always wear _____ protection. Hearing protection is also _____.

b. Never try to clear a _____ nailer while it is still connected to the air supply.

c. Never fire a nailer unless the _____ is in contact with the workpiece.

d. Never carry a nailer with your finger on the _____.

e. Never use _____ gas to power the tool.

f. Air-driven nails sometimes _____.

g. Never operate a nailer at a _____ higher than it was designed to handle.

h. If you are using a belt-driven compressor, make sure that the belts are protected by a _____.

i. Before transporting a compressor, release the _____ in the tank.

j. The _____ connected to a nailer should be in good condition.

k. Make sure the nailer is pointing _____ when you connect a pressurized air hose to it.

2. A pneumatic nailer or _____ is a tool that uses compressed air to drive fasteners into wood.

3. The two basic types of nailers are the _____ fed and the _____.

4. Coil nailers are more common on job sites than strip nailers.

5. A strip nailer can be used longer without the need for reloading than a coil nailer. (T or F)

6. Strip nailers are not easier to fit into tight places. (T or F)

(Continued on next page)

a. _____

b. _____

c. _____

d. _____

e. _____

f. _____

7. Name the parts of the nailer shown in Fig. 14-1.

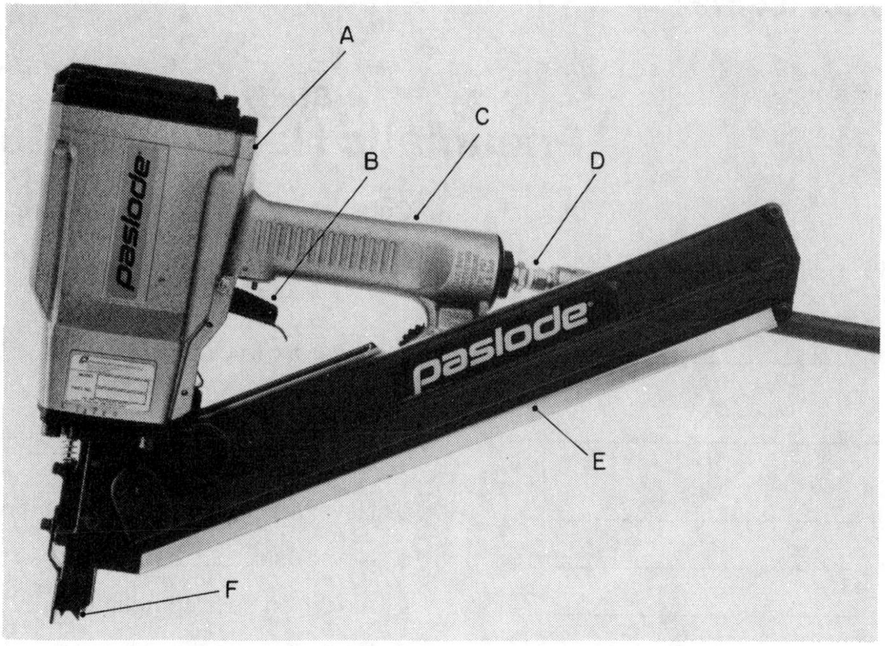

Fig. 14-1.

_____

_____

_____

a. _____

b. _____

_____

_____

_____

_____

_____

_____

_____

_____

8. Because of the narrow shape of the tool, coil nailers fit more easily into tight places. (T or F)

9. All nailers and staplers employ a two-step _____ sequence.

10. Nailers and staplers are attached to a hose that is linked to a _____.

11. The following two things must occur before a tool will fire:

   a. The _____ must be pulled.

   b. The _____ of the tool must be pressed against the workpiece.

12. Most nailers operate on about _____ to _____ psi.

13. A no-compressor nailer has a small internal-combustion engine located in the head of the tool. (T or F)

14. Staplers are used primarily for installing (one wrong): a. sheathing; b. sub-flooring; c. cabinets; d. roofing.

15. Staples should be countersunk into the surface of the wood. (T or F)

16. The crown of the staple should be kept perpendicular to the grain of the wood. (T or F)

17. The crown width for staples used to install roofing must be at least _____ inch.

18. Most building codes accept roofing staples on a one-for-one basis. (T or F)

19. Staples used to install asphalt or fiberglass roofing must be _____.

_____ 20. Staples for roofing should be placed _____ inch above the shingle cutout.

_____ 21. To maintain nailers and staplers in good condition, the tools must be properly _____.

_____ 22. Spray the _____ of the tool with WD-40. Then wipe clean.

_____ 23. Most compressors used on residential job sites are _____ units.

_____ 24. A valve that controls the amount of air pressure reaching a nailer is called a _____.

_____ 25. A line-pressure _____ monitors the pressure in the hose leading to the air tool.

_____ 26. If a worker can fire fasteners faster than a compressor can supply air, the compressor is _____.

_____ 27. An air hose should have a minimum inside diameter of _____ inch.

_____ 28. Long lengths of air hose can be a tripping hazard. (T or F)

_____ 29. Air nailers are loaded with either strips or _____ of nails.

_____ 30. Three types of shanks on nails are (one wrong): a. smooth; b. knurled; c. screw; d. ring.

_____ 31. Nails range in length from _____ inch brads to _____ inch spikes.

_____ 32. Staples are available from _____ inch long to _____ inches long.

_____ 33. Staples can be made from (one wrong): a. steel; b. aluminum; c. copper; d. bronze.

_____ 34. In framing, a 14-gauge 3-inch staple can be used instead of a _____ nail.

_____ 35. Staples used to install wall sheathing should penetrate the framing at least _____ inch.

_____ 36. Pneumatic tools can be used to install corrugated fasteners. (T or F)

_____ 37. Staples used for framing must have a crown at least _____ inch wide.

_____ 38. Staples used for sheathing that is 1/2" thick must be at least _____ inches long.

_____ 39. Corrugated fasteners are excellent for assembling picture frames. (T or F)

_____ 40. The inside diameter for hoses 51 to 100 feet in length should be _____ inch.

To be used with Unit 17

Pages 174-181

Name ______________________

Score: (51 possible) ______

Study Unit 15
Routers

a ______________________
b. ______________________
c. ______________________
d. ______________________
e. ______________________

1. When using the router:
   a. Keep both ____________ on the handles.
   b. Feed in the ____________ direction.
   c. Always lay the router down with the cutter pointing ____________ from you.
   d. Always hold onto the router when it is ____________.
   e. Be certain the power switch is ____________ before connecting the power plug.

______________________ 2. Routing is similar to ____________ except that the motor and cutters are just opposite.

______________________ 3. The portable router consists of a motor with a ____________ attached to the spindle.

______________________ 4. To do straight routing, a ____________ is needed.

______________________ 5. A router can be used to cut a rabbet. (T or F)

______________________ 6. The shape of a decorative edge can be changed by moving the ____________ up or down in the base.

7. Name the parts of the router as shown in Fig. 15-1.

a. ______________________
b. ______________________
c. ______________________
d. ______________________
e. ______________________
f. ______________________
g. ______________________
h. ______________________
i. ______________________
j. ______________________
k. ______________________
l. ______________________
m. ______________________
n. ______________________

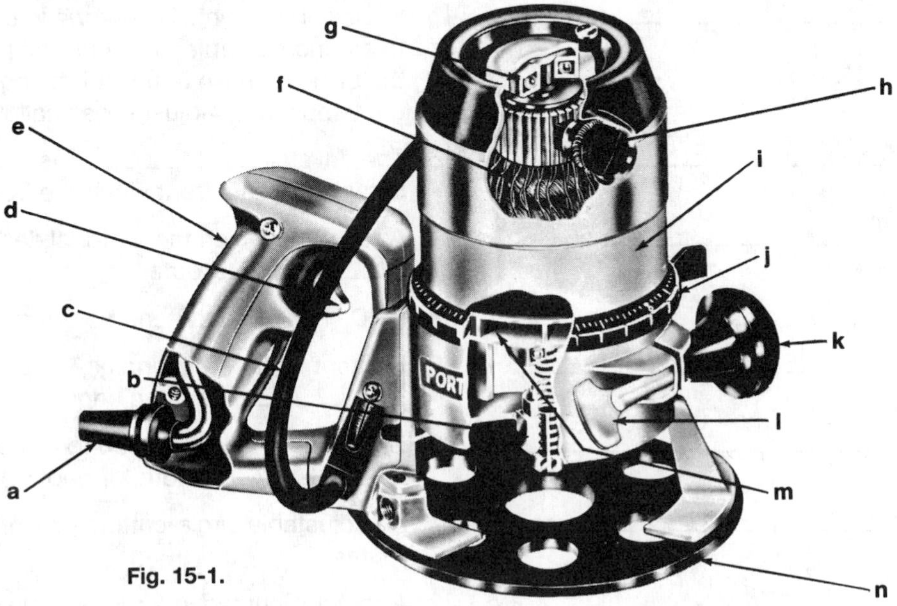

Fig. 15-1.

1 1/4 HP ROUTER

(Continued on next page)

a. _____

b. _____

c. _____

d. _____

e. _____

f. _____

g. _____

h. _____

i. _____

j. _____

k. _____

l. _____

m. _____

_____

_____

_____

_____

_____

_____

_____

_____

_____

_____

_____

_____

8. Name the router bits shown in Fig. 15-2.

Fig. 15-2.

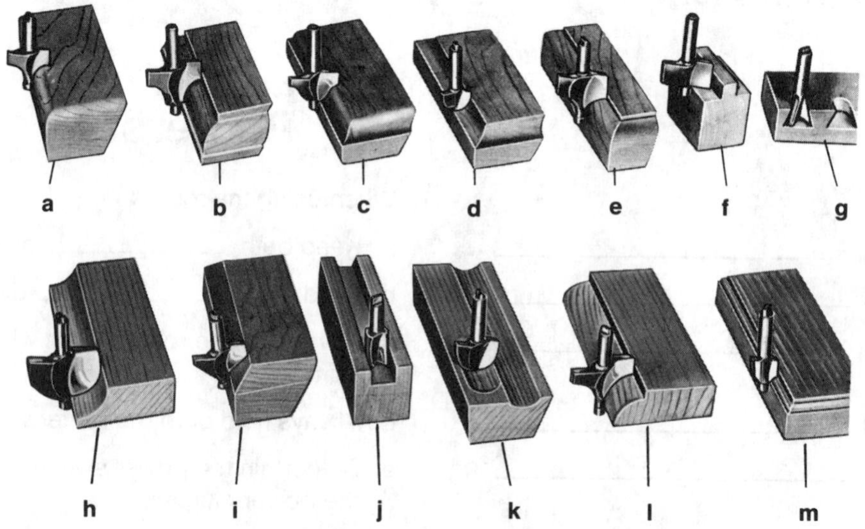

a   b   c   d   e   f   g

h   i   j   k   l   m

9. Many kinds of decorative edges can be cut with a portable router. (T or F)

10. When making a cut on a straight edge, feed from right to left. (T or F)

11. When cutting on circular stock, feed in a _____ direction.

12. All router work is done freehand. (T or F)

13. The joint used for good drawer construction is called the _____ joint.

14. The dovetail joint can be cut with a _____ and dovetail attachment.

15. In making a dovetail joint (one wrong): a. clamp the dovetail attachment to a bench or table; b. clamp the project face side in against the front of the base; c. make a trial cut, being sure that the template guide follows the template; d. adjust for a shallower cut if the trail joint is too loose.

16. The dovetail _____ is a good joint to use for fastening a drawer side to a front when the front "lips over" the cabinet sides.

17. A router on which the motor slides up and down on posts is called a _____ router.

18. A "D" handle router can be operated with one hand. (T or F)

19. Router bits have either high-speed steel or _____- _____ cutting edges.

20. Ball-bearing collars added to router bits can be used to change the _____, width, or depth of the router bit.

21. An adjustable circle-cutting jig works like a _____ to guide the router.

22. A router mounted in a router table operates much like a _____.

To be used with Unit 18

Pages 182-185

Name ______________________

Score: (28 possible) ______

Study Unit 16
Saber Saw

1. Follow these safety rules:
   a. Select the correct __________ for the work. — a. __________
   b. Make sure the work is properly __________. — b. __________
   c. Keep the cutting __________ constant. — c. __________
   d. Hold the __________ securely on the work. — d. __________
2. In making a heavy cut on a piece of 2″ × 4″ at a 45-degree angle, use a blade with __________ teeth per inch. __________
3. At least __________ teeth should be on the cutting surface at all times. __________
4. All saber saws are of the same design. (T or F) __________
5. Name the parts of the hand jigsaw shown in Fig. 16-1.
   a. __________
   b. __________
   c. __________
   d. __________
   e. __________
   f. __________
   g. __________
   h. __________
   i. __________

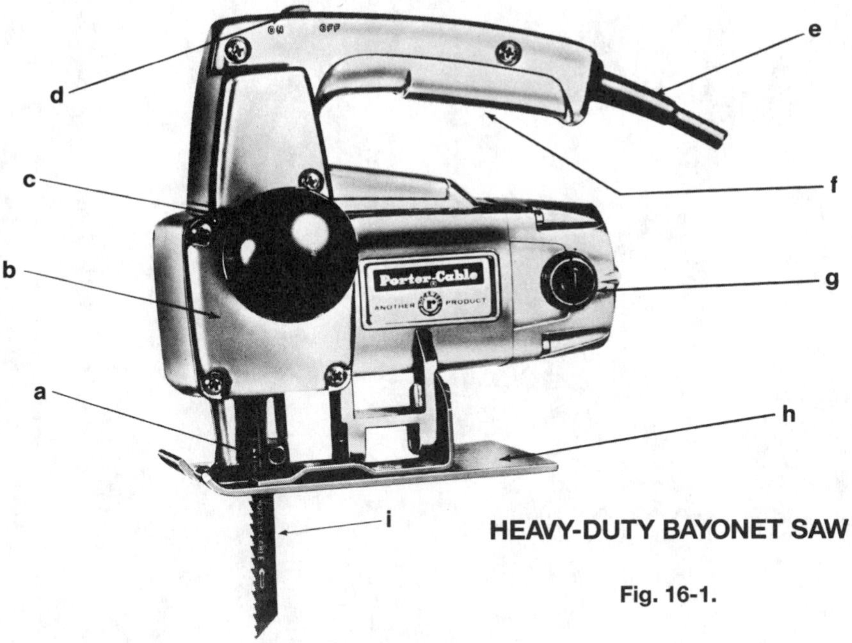

HEAVY-DUTY BAYONET SAW

Fig. 16-1.

6. The hand jigsaw is used primarily for straight and __________ cuts on the job. __________
7. Other names for the portable jigsaw are the __________ or bayonet saw. __________
8. Straight cutting can be done freehand or with the use of a __________ fence or guide. __________
9. Circles can be cut using the ripping __________ as a radius arm. __________

(Continued on next page)

_____   10. The correct blade to use for sawing cardboard is a _____ blade.

_____   11. Making an inside cut without first drilling a hole is called _____ cutting.

_____   12. For cutting plywood, a blade with the following number of teeth should be used *(one right)*: a. 6; b. 10; c. 12; d. 14.

_____   13. The hand jigsaw changes rotary action to up-and-down action. (T or F)

_____   14. A reciprocating saw operates with a back-and-forth movement. (T or F)

_____   15. A reciprocating saw is used mostly for _____ or cabinetwork.

_____   16. When a reciprocating saw is used for remodeling, use a blade that will cut metal. (T or F)

_____   17. Blades for reciprocating saws are available that cut wood, plastic, metal, and _____ rod.

Study Unit 17
Portable Sander

a. _____

b. _____

c. _____

d. _____

e. _____

f. _____

g. _____

h. _____

i. _____

j. _____

k. _____

l. _____

m. _____

n. _____

a. _____

b. _____

c. _____

_____

_____

_____

1. Name the parts of the portable belt sander shown in Fig. 17-1.

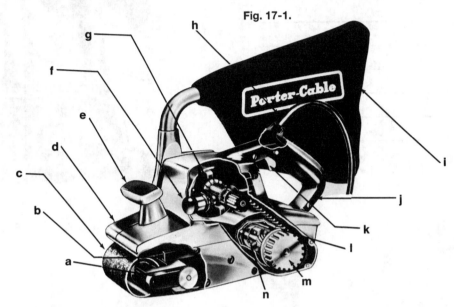

Fig. 17-1.

2. Identify the three principal methods of sanding with the finishing sander shown in Fig. 17-2.

3. Many cutting operations should be done after sanding. (T or F)

4. To use a portable sander (one wrong): a. put the cord over your right shoulder; b. hold the machine with your right hand; c. slowly lower the back of the sander onto the work; d. move the machine forward and back.

5. Finishing or pad sanders operate on one of three basic principles as shown in Fig. 17-2. Which action is least likely to leave cross-grain scratches: a, b, or c?

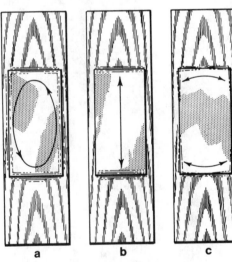

Fig. 17-2.

(Continued on next page)

a. _____

b. _____

c. _____

d. _____

e. _____

f. _____

g. _____

h. _____

i. _____

j. _____

k. _____

6. The finishing sander requires a great deal of hand pressure to do a good job. (T or F)

7. Name the parts of the finishing sander shown in Fig. 17-3.

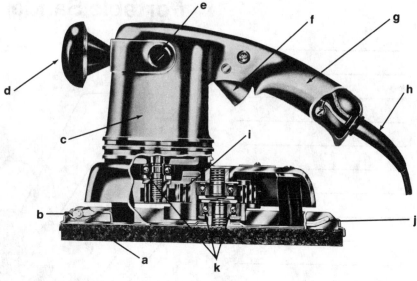

Fig. 17-3.

Study Unit 18
Jointers, Jobsite Planers, and Electric Planes

a. _____

b. _____

c. _____

d. _____

e. _____

f. _____

g. _____

h. _____

i. _____

j. _____

k. _____

1. Name the jointer parts shown in Fig. 18-1.

Fig. 18-1.

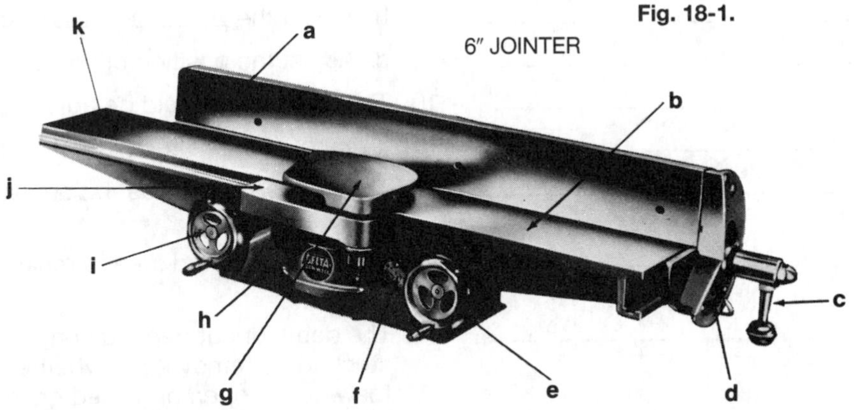

6" JOINTER

2. For the jointer to be safe to use, the following must be done:

a. _____

b. _____

c. _____

d. _____

e. _____

f. _____

g. _____

h. _____

i. _____

_____

_____

_____

_____

a. Keep the knives _____.

b. Be sure the fence is _____.

c. The guard must be in _____ and operating easily.

d. The machine should be at _____ speed before it is used.

e. Never stand _____ the machine.

f. The wood should be cut _____ the grain.

g. Keep the left hand back from the _____ end of the board when feeding.

h. Never apply pressure to the board with your hand directly over the _____.

i. Use good _____ about when stock is too thin or short to joint safely.

3. Operations that can be performed on the jointer include *(one wrong)*: a. planing a surface; b. cutting a rabbet; c. cutting a groove; d. cutting a bevel.

4. The size of a jointer is indicated by the _____ of the knives.

5. A jointer should operate at a speed of about _____ rpm.

6. Another name for the outfeed table is _____ table.

7. Some jointers are made with one table that is adjustable and one table that is fixed. (T or F)

(Continued on next page)

8. Match the descriptions in the left column with the parts in the right column:

a. _____
b. _____
c. _____
d. _____

a. has three or four blades   1. tables
b. covers the cutterhead   2. cutterhead
c. infeed and outfeed   3. fence
d. provides support for the work   4. guard

9. The following adjustments must be made before work can begin:

a. _____
b. _____
c. _____

a. Align and adjust the _____ table.

b. Adjust the _____ table.

c. Adjust the position of the _____.

_____

10. The rear table should be adjusted for each different type of cut. (T or F)

_____

11. If the rear table is too high, the board will be slightly tapered. (T or F)

_____

12. The outfeed table must be adjusted to change the depth of cut on a jointer. (T or F)

_____

13. It is usually a good idea to make a trial cut before beginning work. (T or F)

_____

14. The depth of cut depends on the following (one wrong): a. amount of stock to be removed; b. whether the wood is soft or hard; c. whether the wood is open or closed grain; d. the smoothness of the surface desired.

_____

15. The position of the fence should always remain the same. (T or F)

_____

16. If a board has warp or wind, the cut must be made in a different way. (T or F)

_____

17. You always push the board across the jointer with the push stick or push _____ .

_____

18. The size of the jointer is indicated by the length of the knives. (T or F)

_____

19. When planing end grain, make a short cut at one end and then reverse the stock and feed from the opposite direction. (T or F)

_____

20. In operating the jointer, never adjust the fence while the jointer is running. (T or F)

_____

21. The jointer is a good machine to use when cutting a _____ with the grain.

_____

22. The planer cannot be used to square up stock for stair balusters. (T or F)

_____

23. The width of the rabbet to be cut is controlled by the position of the (one right): a. outfeed table; b. guard; c. fence; d. infeed table.

_____

24. The depth of the rabbet to be cut is controlled by the position of the (one right): a. outfeed table; b. guard; c. fence; d. infeed table.

_____

25. The purpose of a jointer is to reduce the thickness of a board, smooth its surface, and make one side parallel to the other. (T or F)

_____

26. The depth of cut of a jobsite planer should be set (one right): a. while the machine is running; b. at any time; c. while the stock is being machined; d. when the machine is at a dead stop.

_____

27. The portable electric plane has a cylindrical cutterhead. (T or F)

_____

28. The depth of cut on an electric hand plane can be varied. (T or F)

_____

29. A portable electric plane should be guided using both hands. (T or F)

_____

30. the electric hand plane can make bevel cuts. (T or F)

Name _____

Score: (43 possible) _____

Study Unit 19
Plate Joiners

1. Follow these general safety rules when using a plate joiner:

a. _____

b. _____

c. _____

d. _____

e. _____

f. _____

g. _____

h. _____

i. _____

j. _____

k. _____

_____

_____

_____

a. _____

b. _____

c. _____

d. _____

e. _____

f. _____

g. _____

h. _____

a. Several practice cuts should be made on stock to _____ the user with the machine.

b. A plate joiner ejects dust and _____ at a high rate of speed.

c. Never cut a slot in a piece of wood that is _____ held.

d. Wear _____ protection because the plate joiner is noisy.

e. Failure to retract the blade fully may cause _____.

f. Clamp the _____ that is to be cut.

g. Chips and dust from the machine may cause a _____ hazard.

h. Do not disable the _____ points on the faceplate.

i. When changing the blades _____ the power cord.

j. Make sure the blades are _____.

k. Check the operation of the _____ base.

2. Another name for wood splines called plates is _____.

3. Biscuits strengthen the joint and help to _____ the pieces accurately.

4. The plate joiner cannot be used to install butt-joining custom wood flooring. (T or F)

5. Name the parts of the plate joiner shown in Fig. 19-1.

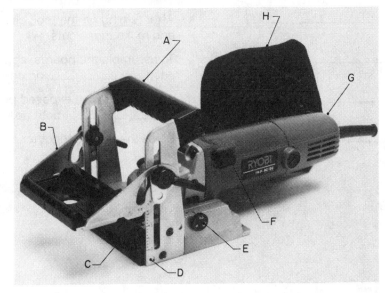

Fig. 19-1.

(Continued on next page)

_____

_____

_____

_____

_____

_____

6. A plate joiner cuts _____ shaped grooves in the edge of wood.

7. The plates are die-cut from _____ (wood) blanks.

8. Plates or biscuits come in _____ standard sizes.

9. A number-20 plate or biscuit is about _____ inch wide and _____ inches long.

10. Plates made of _____ are used for joining synthetic counter-top materials.

11. Wood plates absorb moisture from adhesives. (T or F)

12. Follow these steps in edge-joining two pieces of 1″ × 6″ board for shelves:

a. _____   a. Place the boards edge to _____.

b. _____   b. Mark a line across the _____ with a pencil.

c. _____   c. These marks should be 8 to _____ inches on center.

d. _____   d. Adjust the joiner's depth of cut for the _____ of the plate.

e. _____   e. Adjust the fence to _____ the cut in the thickness of the board.

f. _____   f. Clamp _____ board at a time to the workbench.

g. _____   g. Press the faceplate of the machine against the stock, using the centerline _____ on the tool to align it with the layout marks.

h. _____   h. Turn the joiner on, and _____ it into the board.

i. _____   i. After the cut is complete, pull the joiner _____ from the stock.

j. _____   j. Line up the joiner with the next _____ line. Continue to make cuts in this manner.

k. _____   k. After turning off the tool, clamp the _____ piece in place and make those cuts.

l. _____   l. To assemble the boards, apply a light coat of glue to the inside of each _____ on one board. Then insert the plates.

m. _____   m. Apply glue to the exposed portions of the plates and the _____ of both boards.

n. _____   n. Press the boards together and _____ them.

Study Unit 20
Scaffolds and Ladders

a. _____2_____
b. _____4_____
c. _____5_____
d. _____1_____
e. _____6_____
f. _____3_____

_____F_____
_____duplex_____

a. _____level_____
b. _____rails_____
c. _____braces_____
d. _____stairs_____
e. _____guying_____
f. _____base plate (ground)_____
_____a-wrong_____

a. _____horizontal_____
b. _____nuts, bolts & fastenings_____
c. _____
d. _____more_____
e. _____electricity_____
f. _____doors or openings_____
_____3_____
_____b_____

1. Match the items at the left to the descriptions at the right:
   a. bracket
   b. manufactured scaffolding
   c. ladders
   d. wood scaffolding
   e. ladder jacks
   f. trestle

   1. constructed on the job to aid workers
   2. support for scaffold parts
   3. a pair of jacks that support scaffolding
   4. can be set up or taken apart at building site
   5. usually made of wood or aluminum
   6. can be attached to a ladder to hold a scaffold plank

2. All scaffolding should be freestanding. (T or F)

3. A _____ head nail is used when constructing wood scaffolding.

4. Scaffolding should have the following characteristics:
   a. Be plumb and _____.
   b. Be provided with proper guard _____.
   c. Have all _____ fastened securely.
   d. Should be climbed by means of _____ or fixed ladders.
   e. If freestanding, should be fastened by _____.
   f. Should rest firmly on the _____.

5. The following are kinds of ladders *(one wrong)*: a. structural; b. straight; c. folding; d. extension.

6. Complete these statements on ladder safety:
   a. Never use ladders in a _____ position.
   b. Be sure that all _____ are tight.
   c. Check to see that _____ have a firm base.
   d. It is better to _____ the ladder than to lean far out to work.
   e. Remember that a metal ladder will conduct _____.
   f. Do not place a ladder in front of _____.

7. When working on a roof, be sure the ladder extends above the roof edge at least _____ feet.

8. An extension ladder should be leaned against a building at an angle of *(one right)*: a. 85°; b. 75°; c. 80°; d. 70°.

(Continued on next page)

_____T_____   9. OSHA has developed regulations regarding safe construction techniques. (T or F)

_____T_____   10. Improperly erected scaffolding is one of the leading causes of accidents in the construction industry. (T or F)

_____duplex_____   11. When constructing wood scaffolding, use _____ head nails.

_____planks._____   12. Horizontal wood pieces on which you stand are scaffolding _____.

_____F_____   13. Lumber plank is better than aluminum plank. (T or F)

_____T_____   14. Laminated wood planks are made specifically for scaffolding. (T or F)

To be used with Unit 23 Name ______________________

Pages 214-224 **Score: (25 possible)** ______

Study Unit 21

Locating the House on the Building Site

Reference 1. One way to locate a house on a lot is to work from a ____________ point or line that can be identified.

true 2. A street is a good reference point. (T or F)

2 3. There are ____________ ways to locate a structure on a lot. (T or F)

c. street light 4. Some reference points for laying out a site are *(one wrong)*: a. sidewalk; b. foundation of nearby building; c. a street light; d. a curbing.

fire hydrant is good

optical (builders) 5. One instrument that can be used to locate a structure on a lot is the ____________ level, also called the dumpy level. horizontal only

transit 6. In addition to the dumpy level, the ____________ level can be used to locate a structure on a lot.

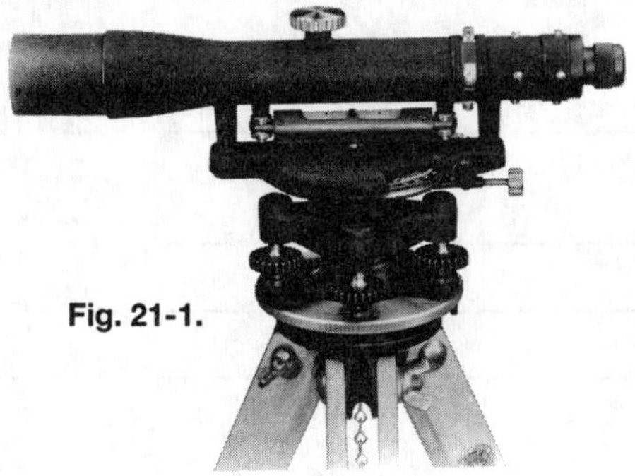

Fig. 21-1.

F 7. A reference point is not needed when using one of the levels in questions 5 and 6. (T or F)

Optical 8. The instrument shown in Fig. 21-1 is an ____________ level.

transit 9. The instrument shown in Fig. 21-2 is a ____________ level.

Optical 10. The instrument used for measuring horizontal angles only is the ____________ level.

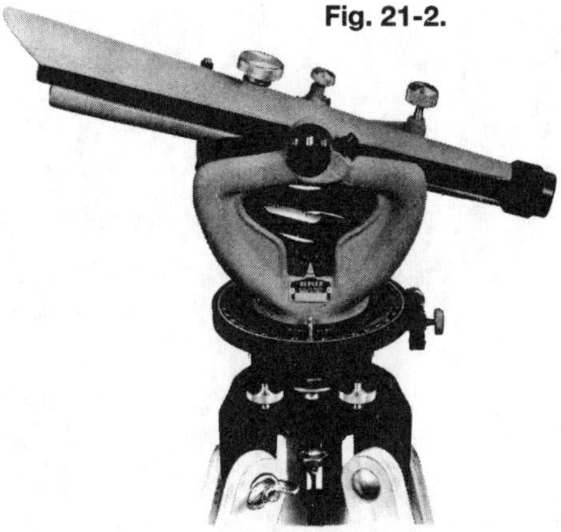

Fig. 21-2.

elevation 11. By careful use of the level, one can see the difference in ____________ between two points.

T 12. To lay out an irregularly shaped structure on a lot, more points are needed than for a rectangular shape. (T or F)

Grade 13. A reference point which refers to the level of the ground where it will touch the foundation of the completed building is called the ____________ line.

(Continued on next page)

_____ Station _____ 14. When setting up a level, the point over which the level is directly centered is called the _____ mark.

_____ F _____ 15. Once a level has been set up, it can be easily moved to adjust the position of the tripod. (T or F)

_____ d _____ 16. After the corners of the house have been located on the lot, the outline of the house is marked with *(one right)*: a. stakes; b. a chalk line; c. metal posts; d. batter boards.

_____ finish _____ 17. The ground level of a lot after the lot has been graded is called the _____ grade.

_____ a wrong _____ 18. Foundation walls should extend above this ground level *(one wrong)*: a. for better appearance; b. to protect the wood finish of the house; c. to protect the house frame; d. to keep the house above the grass line.

_____ a wrong _____ 19. Test borings of subsoil should be taken to determine *(one wrong)*: a. if oil is present; b. the height of the water table; c. if there is a rock ledge; d. the kind of soil.

_____ F _____ 20. Footings and basement should be excavated at the same time. (T or F)

_____ F _____ 21. Basement excavations should always have vertical walls. (T or F)

_____ F _____ 22. The outside of the foundation wall is the line for the basement excavation. (T or F)

_____ level _____ 23. The telescope of a _____ is fixed in position. *for vertical grade*

_____ laser _____ 24. A single crew member can use a _____ level.

_____ angles _____ 25. One type of transit, called an electronic transit, reads _____ electronically.

Study Unit 22
Concrete and Footings

_____ C _____ ✓ 1. Concrete contains *(one wrong)*: a. gravel; b. water; c. a chemical hardening agent; d. sand.

_____ hydration _____ ✓ 2. When water is added to concrete ingredients, a reaction called _____ takes place.

_____ harden _____ ✓ 3. This reaction causes concrete to _____.

_____ T _____ ✓ 4. Water used for making concrete should be clean enough to drink. (T or F)

_____ F _____ ✓ 5. Clay mixed with gravel helps to strengthen the concrete. (T or F)

_____ F _____ ✓ 6. The proportions for mixing concrete are relatively unimportant. (T or F)

7. A 1:2:3 batch of concrete has the following proportions:

a. _____ ✓ a. One part _portland cement_

b. _____ ✓ b. Two parts _sand_ .

c. _____ ✓ c. Three parts _gravel_ or. crushed stone

_____ T _____ ✓ 8. Whenever possible, pour concrete continuously. (T or F)

_____ T _____ ✓ 9. The area where concrete is being poured should be kept level. (T or F)

_____ vibrate _____ 10. Spade or _____ the concrete to remove air pockets.

_____ F _____ ✓ 11. Concrete should be allowed to dry rapidly. (T or F)

_____ F _____ ✓ 12. The faster concrete dries, the stronger it is. (T or F)

_____ T _____ ✓ 13. Concrete sets at a rate directly related to temperature. (T or F)

_____ foundations _____ ✓ 14. Footings are the _foundation_ of a structure.

_____ F _____ ✓ 15. Footings are a standard size regardless of climate or soil condition. (T or F)

_____ T _____ ✓ 16. The depth of footings varies with the climate. (T or F)

_____ T _____ ✓ 17. The best footings for homes are made of concrete. (T or F)

_____ a _____ ✓ 18. Footings should be at least *(one right)*: a. 6″ thick; b. 7″ thick; c. 5″ thick; d. 8″ thick.

19. Two rules for footings are:

a. _____ depth _____ ✓ a. Footing _depth_ should equal wall thickness.

b. _____ ½ _____ ✓ b. Footings should project out from sidewalls _____ the thickness of the walls.

_____ T _____ ✓ 20. Footings should go below the frost line. (T or F)

_____ F _____ ✓ 21. Soil condition has little to do with footing size. (T or F)

(Continued on next page)

_____d_____ ✓ 22. Footings are required for *(one wrong)*: a. chimneys; b. fireplaces; c. furnaces; d. sidewalks.

_____30 × 30 × 12_____ ✓ 23. Piers, posts, and columns are usually made in two sizes: 24″ × 24″ × 12″ and _____.

_____steps_____ ✓ 24. When a lot is steep, the footings are poured in the shape of _____.

_____drains_____ ✓ 25. To help keep moisture out of a home, foundation or footing _____ are installed.

26. Match the admixtures on the left with the descriptions on the right.

a. _____4_____ ✓
b. _____1_____ ✓
c. _____5_____ ✓
d. _____3_____ ✓
e. _____2_____ ✓

a. air-entraining
b. retarding
c. accelerating
d. water-reducing
e. superplasticizing

1. concrete sets up slowly
2. can do one of two things
3. makes the concrete stronger
4. introduces bubbles in the concrete
5. concrete can be put in service quickly

Study Unit 23
Poured Concrete Foundation Walls

Regulations ✓ 1. In addition to soil conditions, building _Regulations_ must also be considered when planning construction of foundation walls.

_T_ ✓ 2. Foundation walls bear the weight of an entire house. (T or F)

_a_ ✓ 3. Basement wall height should be at least (one right): a. 7'; b. 6 1/2'; c. 8'; d. 8 1/2'.

_T_ ✓ 4. For a poured concrete wall, each side of the wall must have a wood form. (T or F)

double formed ✓ 5. This wood form is called _double formed_.

_F_ ✓ 6. Forms can be used only once. (T or F)

_F_ ✓ 7. If the forms are straight and plumb, bracing is unnecessary. (T or F)

_d_ ✓ 8. During good weather, poured foundation walls dry in not less than (one right): a. 24 hours; b. three days; c. a week; d. two days.

damp proofed ✓ 9. To keep poured concrete walls from rain seepage, they should be _or moisture proof_

asphalt ✓ 10. Concrete walls are protected from rain seepage by an application of hot tar or _____.

_T_ ✓ 11. A house with a basement is more expensive than one built with a crawl space. (T or F)

piers (girder) ✓ 12. In a house with a crawl space, the floor beams are supported by _piers_ made of concrete or concrete block.

_b_ ✓ 13. These should extend above the ground line at least (one right): a. 10"; b. 12"; c. 14"; d. 16".

_T_ ✓ 14. The size of exterior wall piers depends on the weight of the house. (T or F)

_F_ ✓ 15. The brick should be completely sealed with mortar. (T or F)

or straping system _anchor bolt_ ✓ 16. The device which gives strength and stability to the sill plate is the _____. _15" for masonry_ _8" for concrete_

_T_ ✓ 17. The device named in question 16 must be imbedded deeper in concrete block than in poured concrete walls. (T or F)

_8'_ ✓ 18. Anchor bolts should be placed _____ feet apart. _NC = 6' apart_

lintels ✓ 19. Poured concrete walls must have reinforcements over doors and windows called _____.

ties ✓ 20. Wood-frame walls covered with a masonry veneer such as brick must be tied to the veneer with metal _____. _avg 3½ sq feet +_ _24"_

_T_ ✓ 21. Sheathing should be applied between brick and wood-frame walls. _horizontally_ (T or F)

(Continued on next page)

a. _____Sub floor_____ ✓
b. _____joist_____ ✓
c. _____sill plate_____ ✓
d. _____anchor bolt_____ ✓
e. _____sill sealer_____ ✓
f. _____foundation wall_____ ✓

p. 239
in text book

g = header
header joist
or box sill

22. Name the construction parts shown in Fig. 23-1.

Fig. 23-1.

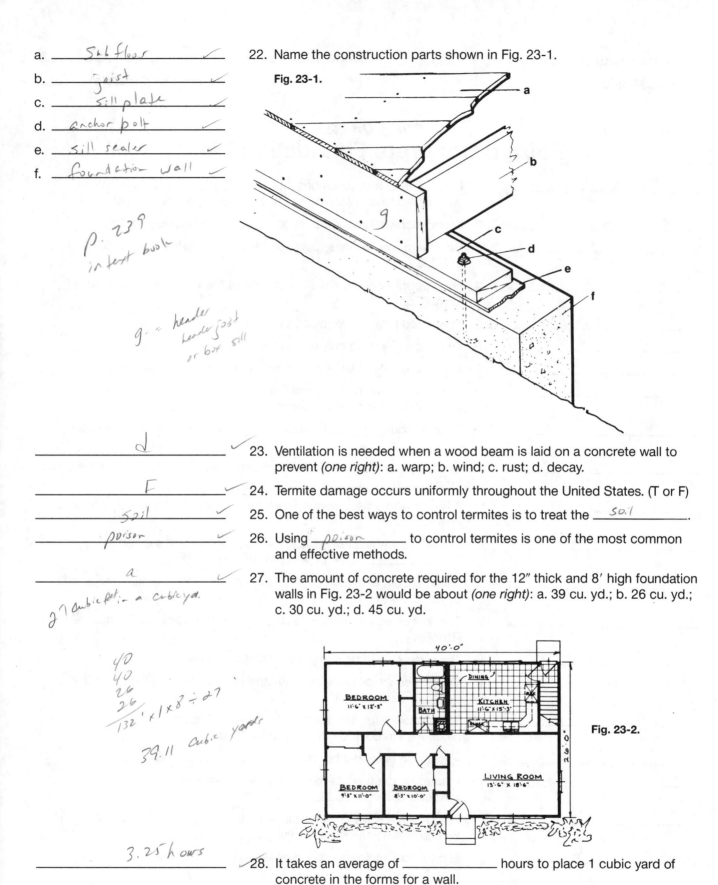

_____d_____ ✓
23. Ventilation is needed when a wood beam is laid on a concrete wall to prevent *(one right)*: a. warp; b. wind; c. rust; d. decay.

_____F_____ ✓
24. Termite damage occurs uniformly throughout the United States. (T or F)

_____soil_____ ✓
25. One of the best ways to control termites is to treat the ___soil___.

_____poison_____ ✓
26. Using ___poison___ to control termites is one of the most common and effective methods.

_____a_____ ✓
27. The amount of concrete required for the 12″ thick and 8′ high foundation walls in Fig. 23-2 would be about *(one right)*: a. 39 cu. yd.; b. 26 cu. yd.; c. 30 cu. yd.; d. 45 cu. yd.

27 cubic feet in a cubic yd.

40
40
26
26
───── 132' × 1 × 8 ÷ 27
132

39.11 cubic yards

Fig. 23-2.

3.25 hours
28. It takes an average of _____ hours to place 1 cubic yard of concrete in the forms for a wall.

Study Unit 24
Concrete Block Foundation Walls

_____ 15 5/8" _____   1. Modular concrete blocks are usually 7 5/8 inches high and _____ inches long.

_____ F _____   2. Concrete block comes in one width or thickness. (T or F)

_____ Course _____   3. A series of building materials installed one layer over the other is called a _____. *a course is a row & a series of Rows is a bond*

_____ 7'4" _____   4. The height of a basement wall made of concrete blocks is about _____ as measured from the joists to the basement floor.

_____ 11 _____   5. The number of courses customarily laid to build a concrete block wall for a basement is _____.

_____ F _____   6. Laying concrete block requires the use of forms. (T or F)

_____ b _____   7. The mortar joints between concrete blocks should measure in width about *(one right)*: a. 1/4"; b. 3/8"; c. 1/3"; d. 1/2".

_____ pilaster _____   8. A concrete block wall may be reinforced by a column of blocks called a _____.

_____ F _____   9. In the stack bond pattern of laying cement block, the center of one block is laid over the joint in the two below. (T or F)

_____ T _____   10. Reinforcement is required for the stack bond pattern. (T or F)

_____ a _____   11. Freezing weather has the following effects on mortar *(one wrong)*: a. faster setting; b. low adhesion; c. less strength; d. failure in joints.

_____ Asphalt _____   12. Waterproofing with a coat of _____ after applying a coating of cement-mortar will normally assure a dry basement.

_____ felt _____   13. Extra waterproofing can be provided by applying roofing _____ or a similar material over the first waterproof coating and covering this with hot tar or asphalt if desired.

14. Mortar strength depends upon:

a. _____ moisture _____   a. The rate at which concrete block absorbs _____ from the mortar.

b. _____ workability _____   b. The plasticity or _____ of the mortar.

c. _____ water _____   c. How much _____ the mortar retains.

d. _____ workmanship _____   d. The quality of _____ displayed by the mason.

_____ C   p.139 _____   15. When air temperature is 80°F. or higher, mortar will remain usable for *(one right)*: a. 1 hour; b. 3 hours; c. 2 1/2 hours; d. 3 1/2 hours.
     masonry book

_____ F _____   16. Concrete block should be watered down before it is used. (T or F)

_____ T _____   17. Several of the first blocks should be positioned before mortar is applied. (T or F)

(Continued on next page)

Corner 18. The _____ block should be the first one laid.

_F_ 19. Proper alignment of the first course of blocks is less important than for succeeding courses. (T or F)

_F_ 20. A larger mortar-bedding area is provided by laying the blocks with the smaller holes down. (T or F)

level 21. To make sure the blocks are laid in an even line, check each block with a _____ or straightedge.

story 22. Use a _____ pole or a course pole to make sure the top of each block is 8″ above the previous one.

closure 23. The final gap in the wall is filled with a _____ block.

_F_ 24. Pressing your thumb into mortar leaves an imprint if the mortar is "thumbprint hard." (T or F)

tooling 25. A process of pressing mortar tightly into the joints on both sides is called _____.

_F_ 26. Tooling should be done while the mortar is still very soft. (T or F)

_d_ 27. A control joint is built into a concrete block wall for the following reasons _(one wrong)_: a. to control cracking due to stress; b. to permit slight movement in the wall; c. to give the wall longer life; d. to make the wall waterproof.

_a_ 28. A control joint _(one wrong)_: a. should be thicker than the mortar joints; b. is built vertically into the wall; c. should be plumb; d. should be caulked if it shows.

_T_ 29. To prevent mortar from smearing on concrete block, remove the mortar spatters after they have dried. (T or F) _s/b F_

_F_ 30. It is not necessary to anchor intersecting concrete block walls. (T or F)

_T_ 31. The tops of windows and doors require lintels. (T or F)

_F_ 32. Fiberglass-reinforced mortar should be mixed for 10 to 15 minutes. (T or F)

joint guide 33. When applying fiberglass-reinforced mortar, a sheet metal device called a _____ guide is used when the application is interrupted for more than an hour.

_T_ 34. The wall should be wetted down prior to fiberglass mortar mix application. (T or F)

_T_ 35. The finished grade should be sloped away from the foundation walls for good surface drainage. (T or F)

_T_ 36. Backfill material should be expected to settle. (T or F)

_F_ 37. Less block is needed per square foot of wall area with fiberglass-reinforced mortar than is needed for conventional mortared block walls. (T or F)

stone 38. Split-face blocks have one rough face that looks something like _____.

_T_ 39. There are several kinds of specialty blocks for specific purposes. (T or F)

Study Unit 25
Slab and Flatwork

_____ 5" _____ 1. Concrete flatwork is usually _____ inches or less in thickness.

_____ T _____ 2. Flatwork construction is the best foundation choice for a house without a basement. (T or F)

_____ d _____ 3. In a combined slab and foundation the footing should extend below the grade line (one right): a. 6"; b. 8"; c. 10"; d. 12".

_____ T _____ 4. The slab requires a vapor barrier. (T or F)

_____ Frost _____ 5. In cold climates, foundations must extend below the _____ line.

_____ c _____ 6. Good vapor barriers include (one wrong): a. heavy plastic film; b. asphalt-laminated sheet; c. gravel; d. heavy roofing.

7. Concrete floor slabs require certain construction details:

a. _____ slab _____ a. Vapor barrier laid under the _____.

b. _____ ground _____ b. Finish floor level above the natural _____ level.

c. _____ topsoil _____ c. Removal of _____.

d. _____ gravel _____ d. After laying of water and sewer lines, a covering of _____ or crushed rock.

e. _____ wall _____ e. A good rigid insulation installed around the outside of the _____.

f. _____ surface _____ f. A smooth, troweled _____.

_____ F _____ 8. No reinforcement is required for concrete floor slabs. (T or F)

_____ compacted _____ 9. A floor slab will not settle uniformly unless the earth below the slab has been _____.

_____ a _____ 10. Before tamping it down, the subgrade should (one wrong): a. be thoroughly dampened; b. have all organic material removed; c. be rough graded; d. be leveled off.

_____ fill _____ 11. After the subgrade has been prepared, a layer of _____ is compacted over it.

_____ c _____ 12. Fill may consist of (one wrong): a. crushed stone; b. coarse slab; c. wood chips; d. gravel.

_____ grout _____ 13. The granular layer is then covered with a stiff coat of _____.

_____ T _____ 14. After applying a dampproofing membrane, the concrete floor slab is poured. (T or F)

_____ d _____ 15. Compacting the cement for the floor slab can be done by (one wrong): a. vibrating; b. spading; c. tamping; d. raking.

(Continued on next page)

_a_   16. Concrete should be allowed to cure *(one wrong)*: a. by natural drying in the air; b. covered with burlap; c. by sprinkling with water to keep the covering wet; d. covered with waterproof paper.

trowel   17. Linoleum and asphalt cannot be installed over concrete unless the concrete surface has been smoothed with a steel _____.

F   18. One disadvantage to using concrete as flooring is that it can't be colored. (T or F)

                  19. Match the items at the left with the descriptions at the right:
a. ___2___   a. porches and basement floors   1. surface should be coarse
b. ___1___   b. driveways   2. surface should have smooth-troweled finish
c. ___3___   c. sidewalks   3. surface should be non-slippery

screeding   20. The process of removing excess concrete is called _____.

level   21. Excess concrete is removed in order to make the surface _____.

Chipping   22. Concrete is sometimes edged to prevent _____ or damage.

joints   23. Random cracks that develop in concrete can be controlled by cutting contraction _____ in the surface.

float   24. The tool shown in Fig. 25-1 is a wood _____.

Fig. 25-1.

floating   25. For a smoother finish, screeding is followed by _____.

moisture   26. If done too soon, floating raises fines and _____ to the surface of concrete.

trowel   27. The tool shown in Fig. 25-2 is a metal _____.

b   28. The tool in Fig. 25-2 makes the concrete surface *(one wrong)*: a. smooth; b. rippled; c. free of marks; d. even.

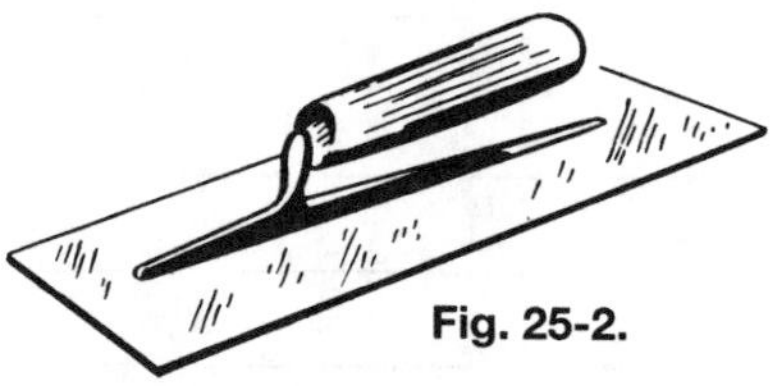

Fig. 25-2.

knee boards   29. The cement finisher can work on large areas without leaving dents by using _____.

100   30. A 12 percent grade in a driveway means that it rises 12 feet in _____ feet.   1' in 8'

Ribbon   31. Two basic types of driveway are the slab and the _____ types.

5   32. Concrete for a driveway should be about ___5___ inches thick.

d   33. Expansion joints in a driveway should be located *(one wrong)*: a. where driveway meets a sidewalk; b. along a curb; c. about 40' apart; d. about 20' apart.

5%   34. Sidewalks that slope should have a grade not greater than _____ percent.   12% in NC

drain   35. The basement floor should be equipped with at least one _____.

F   36. A basement floor requires no screeding. (T or F)

Study Unit 26
Framing Methods

_____ T _____   1. Wood-frame construction is found in most homes on this continent. (T or F)

2. A wood-frame house has the following characteristics:

a. ____ shingles ____   a. Exterior of stucco, brick, wood siding, or wood __shingles__.

b. ____ less ____   b. Costs _____ than other types.

c. ____ better ____   c. Insulating qualities are _____ than other types.

d. ____ Architectral ____   d. Almost any _____ style can be produced.

_____ F _____   3. A disadvantage of a wood-frame house is its lack of durability. (T or F)

Conventional ____ Common ____   4. Platform and balloon are two kinds of _____ framing.

_____ T _____   5. The construction method most commonly found in one-story houses is platform-frame construction. (T or F)

_____ T _____   6. A house frame with lumber and covered with brick or stucco is considered a wood-frame house. (T or F)

_____ d _____   7. In platform-frame construction *(one wrong)*: a. the building process is easier than with balloon frame; b. each floor is constructed separately; c. wall framing can be assembled on the floor and tilted into place; d. the cost is higher than balloon framing.

_____ joists _____   8. In balloon framing, studs and first-floor _____ rest on the foundation.

_____ T _____   9. Balloon framing is less affected by expansion and contraction than platform framing. (T or F)

_____ T _____   10. Post-and-beam construction is not one of the conventional framing methods. (T or F)

_____ b _____   11. Post-and-beam construction *(one wrong)*: a. has interior roof planks exposed; b. has smaller framing members; c. lends itself to striking architectural effects; d. requires less labor.

_____ T _____   12. The framing pieces for metal framing are made in a factory. (T or F)

_____ F _____   13. Paul Revere's house is older than the "House of the Seven Gables." (T or F)

_____ c _____   14. Metal framing has the following advantages *(one wrong)*: a. will not rot or decay; b. does not warp or twist; c. will not expand with temperature; d. will not shrink or swell with changes in temperature.

_____ T _____   15. Metal framing can be produced in lengths not available with wood framing. (T or F)

_____ Runners _____   16. The top and bottom plates of metal framing are called _____.

(Continued on next page)

_____T_____ 17. The screws used to assemble metal framing have Phillips heads. (T or F)

_____F_____ 18. Metal framing for exterior walls is common in home construction. (T or F)

_____T_____ 19. Windows for metal framing are prefabricated and painted. (T or F)

___grommets___ 20. Wiring for metal framing passes through insulated _____.

___foam-core___ 21. Panels of _____ are sometimes used to form the shell of a house.

_____3½"_____ 22. The panels consist of thick rigid insulation that is _____ inch thick sandwiched between sheets of plywood or OSB board.

___beams___ 23. A timber frame is a structural skeleton of posts and _____ connected with wooden joinery.

_____T_____ 24. A timber frame is a freestanding structural system. (T or F)

___pegs___ 25. Interlocking wood joinery for timber framing is secured by wooden _____.

Name _____

Score: (39 possible) _____

Study Unit 27
Floor Framing

1. Match the floor framing parts at the left with the descriptions at the right:

   a. _____4_____ ✓
   b. _____2_____ ✓
   c. _____5_____ ✓
   d. _____3_____ ✓
   e. _____1_____ ✓

   a. joists
   b. posts
   c. sills
   d. girders
   e. subflooring

   1. wood laid over floor joists
   2. pieces of wood or steel which support the girders
   3. large beams which support floor joists
   4. rest on top of girders
   5. anchored to the foundation wall for fastening and supporting the joists

2. Name the parts shown in Fig. 27-1.

   a. ___joist___ ✓
   b. ___sill plate___ ✓
   c. ___wall foundation___ ✓
   d. ___footing___ ✓
   e. ___girder___ ✓
   f. ___Post___ ✓
   g. ___footing___ ✓

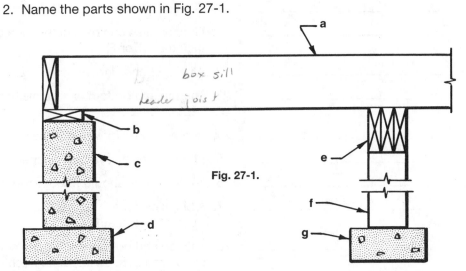

box sill
header joist

Fig. 27-1.

3. There are two types of wood-sill construction used over foundation walls. (T or F) *platform + ballon*

4. Sills should be attached to the wall with 1/2" _____.

   ___T___ ✓
   ___bolts___ ✓
   ___2___ ✓
   ___16"___ ✓
   ___sill___ ✓

5. There should be at least _____ bolts in ~~each pair of~~ sills.

6. Joists are usually placed _____ inches on center.

7. The header joist should be aligned with the outside edge of the _____.

   ___F___ ✓
   ___joists___ ✓
   ___bridging___ ✓
   or wood blocks

8. For openings in floor framing for such things as stairs, chimneys, or fireplaces, the framing members should be tripled. (T or F) *doubled*

9. For a bay window, floor _____ should extend beyond the foundation wall.

10. Beams placed at an angle to joists are called _____.
    90°

(Continued on next page)

T	✓	11. Bridging stiffens the floor and distributes the load evenly on a joist. (T or F)
T	✓	12. Bridging can be either wood or metal. (T or F)
a	✓	13. Girder floor framing is different from conventional floor framing in the following ways (one wrong): a. better for cold climates; b. can be built faster; c. members must be heavier; d. popular for houses with no basement.
joists	✓	14. Subflooring is laid directly over the floor _____.
T	✓	15. Plywood makes an excellent subflooring. (T or F)
use F	✓	16. Floor framing can be held together with construction adhesives. (T or F)
plywood	✓	17. Modern job sites commonly use _____ or OSB for subflooring.
hanger	✓	18. The most common metal connector used in floor framing is the joist _____.
T	✓	19. Floor trusses are made in a factory to job specifications. (T or F)
16"	✓	20. The common depths of floor trusses are 14" and _____".
T	✓	21. Floor trusses can be long enough to reach from one side of the house to the other. (T or F)
3	✓	22. A parallel-chord truss has _____ basic parts.
Chords	✓	23. The top and bottom of a parallel-chord floor truss are called _____.
I joists	✓	24. Another name for wood I-beams is _____.
F	✓	25. I-beams are always built to specific lengths in a factory. (T or F)
F	✓	26. Wood beams should not be stored on edge. (T or F)
web	✓	27. A block called a _____ stiffener should be added to both sides of an I-beam.
metal web	✓	28. The top and bottom chords are connected by devices made of wood or galvanized _____. Each of these connectors is called a _____.

To be used with Unit 30

Pages 316-350

Name ______________________

Score: (43 possible) ______

Study Unit 28
Wall Framing

_____c_____ 1. The outside walls of a house *(one wrong)*: a. support the roof; b. act as framework for attaching exterior facing; c. are the insulation for the house; d. add design to the house's appearance.

_____e_____ 2. The exterior wall contains *(one wrong)*: a. interior and exterior coverings; b. windows and doors; c. insulation; d. studs; e. gussets.

_____T_____ 3. Wall-framing members are called studs. (T or F)

_____warpage_____ 4. Studs must be stiff, good at holding nails, and free from ____________.

_____platform_____ 5. Of the two general types of wall framing, ____________ construction is used the most.

6. Match the items at the left with the descriptions at the right:

a. _____2_____
b. _____3_____
c. _____4_____
d. _____1_____
e. _____5_____
f. _____6_____

a. studs
b. sole plate
c. top plate
d. corner post
e. trimmer studs
f. header

1. forms both inside and outside corner
2. slender wood members placed vertically
3. laid horizontally to carry bottom end of studs
4. connecting link between wall and roof
5. support the header over an opening
6. horizontal member installed over an opening

_____crippled_____ 7. The members installed vertically over and under a window opening are called ____________ studs.

_____c_____ 8. The allowance made for framing in doors and windows is called *(one right)*: a. header; b. top plate; c. rough opening; d. finish opening.

_____trimmer_____ 9. The allowance made for a door or window is the distance between the ____________ studs.

10. Name the parts of the wall framing shown in Fig. 28-1.

a. _____studs_____
b. _____top plate_____
c. _____header_____
d. _____cripple_____
e. _____sill_____
f. _____bottom or sole plate_____

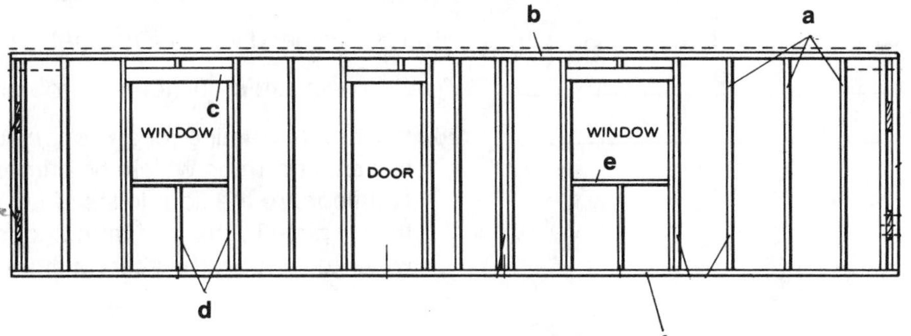

Fig. 28-1.

(Continued on next page)

a. _____ roof _____ ✓
b. _____ rafters _____ ✓
c. _____ top _____ ✓
d. _____ walls _____ ✓

a. _____ floor joists _____ ✓
b. _____ studs _____ ✓
c. _____ foundation _____ ✓

_____ story _____ ✓

_____ T _____ ✓

_____ header _____ ✓

_____ sill _____ ✓

_____ F _____ ✓

_____ T _____ ✓

_____ corners _____ ✓

_____ level _____ ✓

_____ bracing _____ ✓

_____ F _____ ✓

_____ partitions _____ ✓

_____ a _____ ✓

_____ T _____ ✓

_____ soffit _____ ✓

_____ prefabricated _____ ✓

_____ c _____ ✓

11. The top plate does the following:

    a. Acts as a connection between the wall and _____.

    b. Supports the lower ends of the _____.

    c. Ties the studding together at the _____.

    d. Forms a finish for the _____.

12. The location of studs should be determined from a common point because:

    a. Studs will then be directly over floor _____.

    b. Ceiling joists and rafters will be directly over _____.

    c. Members will be aligned from the rafter down to the _____ wall of the building.

13. A full-size layout that shows such things as floor level, ceiling height, and door and window elevations is called a _____ pole.

14. Corner posts should consist of more than one stud. (T or F)

15. When an opening must be cut in the studs for a door or window, a _____ is installed to support the lower ends of the top studs.

16. For a window, the tops of the cut bottom studs are supported by a rough _____.

17. When building the exterior walls, the shorter walls are usually erected first. (T or F)

18. In one method of wall framing, an entire wall can be assembled lying flat on the subfloor and then lifted into place. (T or F)

19. To be sure the wall framing is straight, all exterior and intersecting _____ should first be plumbed.

20. The straightening is accomplished with either a _____ or a plumb bob.

21. Temporary _____ is nailed in place to keep the walls straight.

22. A nonbearing wall is one that supports ceiling joists. (T or F)

23. Walls that divide the inside space of a house are called _____.

24. Special framing is required for openings for (one wrong): a. windows; b. pipes; c. bathtubs; d. heating vents.

25. Special framing requires extra material for strength. (T or F)

26. The bulkhead above a cabinet is called a _____.

27. Cabinets can be built on the job or _____.

28. In platform framing (one wrong): a. the second floor of a multilevel house has a platform on which the carpenter can work; b. ceiling joists for the first floor are the floor joists of the second floor; c. wall studs go from the sill plate to the roof rafters; d. ceiling joists of the first floor must be wider to support the second story.

Study Unit 29
Structural Wall Sheathing

_____ sheathing _____

_____ F _____

_____ T _____

_____ T _____

_____ T _____

_____ gypsum _____

_____ F _____

_____ plywood _____

_____ twice _____

_____ F _____

_____ 4' _____

_____ c _____

_____ square _____

_____ T _____

_____ T _____

_____ plywood _____

_____ paper _____

_____ 2' _____

_____ F _____

a. _____ roof _____

b. _____ subfloor _____

c. _____ wall frames _____

1. The inner layer of the outside wall is called _____.

2. Siding is a structural element of the wall. (T or F)

3. Sheathing is part of the wall framing. (T or F)

4. Insulating sheathing requires diagonal corner bracing. (T or F)

5. Structural sheathing is a structural element of the house. (T or F)

6. Three common kinds of structural sheathing are wood, plywood, and _____.

7. Horizontal sheathing does not require the use of diagonal corner bracing in the wall framework. (T or F)

8. Structural sheathing that covers large areas and adds great strength and rigidity is _____.

9. Plywood sheathed walls are _____ as strong as walls sheathed with diagonal boards.

10. Plywood sheathing must be applied horizontally. (T or F)

11. Plywood sheathing comes in sheets _____' by 8' or longer in length.

12. The following are types of edges on wood sheathing *(one wrong)*: a. shiplap; b. square; c. angle; d. dressed and matched.

13. Wood sheathing must have a _____ edge for butt joinery.

14. Wood sheathing that is dressed and matched has tongue-and-groove joints. (T or F)

15. Oriented-strand board can be used for structural sheathing. (T or F)

16. OSB is generally used as a direct replacement _____.

17. Gypsum sheathing consists of two layers of lightweight _____ with gypsum filler in between.

18. Gypsum sheathing comes in _____' × 8' sheets.

    external
19. Gypsum sheathing should be applied vertically. (T or F)

20. Sheathing can be applied at these different times:

    a. As soon as possible because it adds strength and rigidity to the structure. Walls that have been covered with sheathing provide a more solid structure for the ceiling and _____ members.

    b. When the wall is completely framed and squared and is lying on the _____.

    c. When the _____ frames have been erected, plumbed, and braced.

(Continued on next page)

_____c_____ 21. Carpenters prefer to add sheathing as soon as possible because *(one wrong)*: a. scaffolding is then available for framing the roof; b. the structure is then more rigid; c. the building is warmer to work in; d. there is a more solid structure for ceiling and roof.

_____diagonally_____ 22. Wood sheathing may be applied either horizontally or _____.

_____galvanized_____ 23. Gypsum sheathing is nailed with 1 3/4" or 2" _____ roofing nails.

_____5/16"_____ 24. Plywood sheathing should be a minimum of _____ thick.

_____vertically_____ 25. Plywood is usually applied _____.

_____T_____ 26. Building paper is applied between the sheathing and the siding. (T or F)

_____vapor_____ 27. Building paper should be water-resistant but not _____-resistant.

_____horizontally_____ 28. Building paper should be applied _____.

_____F_____ 29. House wraps are made of paper. (T or F)

_____T_____ 30. House wraps allow water vapor to pass through. (T or F)

_____F_____ 31. House wrap is easy to rip. (T or F)

Study Unit 30
Ceiling Framing

_____ _Truss_ ✓

a. _____ _exterior_ ✓

b. _____ _Gussets_ ✓

c. _____ _king post_ ✓

d. _____ _upper cord_ ✓

e. _____ _lower cord_ ✓

1. Fig. 30-1 is an example of _____ roof construction, one of the basic methods of roof framing for platform-frame construction.

2. The names of the members in Fig. 30-1 are:

Fig. 30-1.

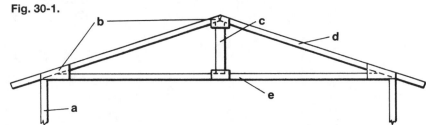

_____ _Conventional_ ✓

_____ _joists_ ✓

_____ _d_ ✓

_____ _T_ ✓

_____ _24_ ✓

_____ _splice_ ✓

_____ _lap_ ✓

_____ _outside_ ✓

_____ _trusses_ ✓
or wall

_____ _beam_ ✓

_____ _strongback_ ✓

3. Besides the type shown in Fig. 30-1, another method of roof framing is called _____ roof construction.

4. The parallel beams that support ceiling loads are called _____.

5. The size of the ceiling joists depends upon *(one wrong)*: a. kind and grade of wood; b. distance they must span; c. load they must carry; d. local supplies available.

6. Ceiling joists form the ceiling of the house and support the ceiling finish. (T or F)

7. Except for the first two ceiling joists, the spacing should be 16 or _____ inches on center.

8. When joining ceiling joists over the partition plate, a plywood or metal joist _____ may be nailed to both sides of the joists.

9. Ceiling joists may also be joined so that they _____ each other.

10. When attaching the ceiling joist to the exterior wall plate, it is important to keep the end of the joist even with the _____ edge of the plate.

11. It is possible for a house to have a large open interior when _____ are used.

12. When ceiling joists and rafters are used, the interior ends of the ceiling joists must be supported by a _____.

13. Ceiling joists with a long span can be given additional support by the construction of a _____.

(Continued on next page)

_____ 672 _____ ✓   14. The board feet of lumber needed for the joists of a house that contains 800 sq. feet of ceiling (joists are 2′ × 8′ and 24′ on center) is _____ board feet.

_____ trusses _____ ✓   15. Prefabricated assemblies placed on framing are called _____.

_____ T _____ ✓   16. Ceiling joists can serve as floor joists for an attic. (T or F)

_____ T _____ ✓   17. The maximum allowable length for ceiling joists that are 2″ × 4″ and spaced 12 inches apart would be 11′ 6″ of Group 1 lumber. (T or F)

_____ 8 _____ ✓   18. The number of board feet in lumber that is 2″ × 4″ and 12′ long is _____ board feet.

_____ 16 _____ ✓   19. The number of board feet in a 2″ × 6″ ceiling joist that is 16′ long is _____ board feet.

Name _____

Score: (35 possible) _____

Study Unit 31
Roof Framing

a. ___gabel + dormer___

b. ___gabel___

c. ___hip___

d. ___gobk___

e. ___gabrel___

f. ___hip + valley___

g. ___shed or lean too___

h. ___flat___

i. ___mansard___

j. ___Dutch hip___

1. Identify the roof styles shown in Fig. 31-1.

a

b

c

d

e

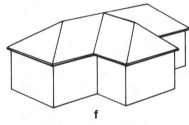
f

g

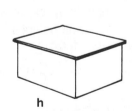
h

i

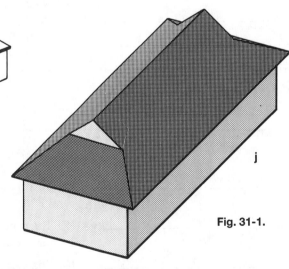
j

Fig. 31-1.

2. A roof should:

a. ___, Weather___

b. ___mantenance___

c. ___wind___

d. ___exterior___

   a. Protect the house in all kinds of ___weather___.

   b. Require a minimum amount of ___mantenance___.

   c. Be strong enough to bear the load of snow and ___wind___.

   d. Be anchored to ___exterior___ walls.

(Continued on next page)

_____F_____ 3. Architectural design bears little relationship to roof style. (T or F)

_____ridge_____ 4. A horizontal piece that connects the upper ends of the rafters is called the _____ board.

_____tail_____ 5. The overhanging part of a rafter is called the _____.

_____slope_____ 6. Another name for the incline of a roof is _____.

7. Match the roof framing terms at the left with the descriptions at the right:

a. _____4_____
b. _____8_____
c. _____5_____
d. _____2_____
e. _____3_____
f. _____7_____
g. _____1_____
h. _____6_____

a. pitch
b. span
c. unit of run
d. cut of a roof
e. measuring line
f. unit rise
g. run
h. rise

1. one-half the distance of the span
2. the rise in inches and the unit of run
3. an imaginary line that runs lengthwise from the outside wall to the ridge
4. the steepness of the roof
5. equal to 1', or 12"
6. vertical distance from the top of the double plate to the upper end of the measuring line
7. number of inches the roof rises per foot of run
8. distance between the outside edge of the double plates

_____Plum_____ 8. A line that runs vertical to a rafter is called a _____ line.

_____level_____ 9. A line that is level with a rafter is called a _____ line.

10. Match the rafters at the left with the descriptions at the right:

a. _____2_____
b. _____1_____
c. _____4_____
d. _____3_____

a. jack
b. common
c. valley
d. hip

1. extend from the plate to the ridge board at 90° to both
2. never extend full distance from the plate to the ridge board
3. extend diagonally from the corners formed by the plate to the ridge board
4. extend diagonally from the plates to the ridge board along lines where two roofs intersect

_____rafters_____ 11. Joists support the floor of a house the same as _____ support the roof.

_____rafters_____ 12. The carpenter should have a roof frame plan to know what kind of _____ are needed to frame the roof.

_____T_____ 13. The exact number of each kind of rafter can be determined from the scale drawing. (T or F)

Study Unit 32
Conventional Roof Framing with Common Rafters

_____F_____ ✓

_____Conventional____ ✓

_____T_____ ✓

_____truss_____ ✓

_____c_____ ✓

_____d_____ ✓

_____joists_____ ✓

_____joists_____ ✓

a. _____ridge_____ ✓

b. _____plumb_____ ✓

c. _____plate_____ ✓

d. _____seat_____ ✓

e. _____plumb_____ ✓

a. _____framing_____ ✓

b. _____pathagorean_____ ✓

c. _____rafter_____ ✓

1. Pitched roofs are built by four methods. (T or F)

2. Figure 32-1 shows the _____ method of roof framing.

3. In the conventional method of roof framing, joists and rafters are put up a piece at a time. (T or F)

4. In _____ roof construction, the roof framing members are usually prefabricated.

Fig. 32-1.

5. Some advantages of the joist and rafter method are (one wrong): a. insulation can be easily installed between joists; b. roof load is carried on the walls; c. it can be done quickly; d. common materials are utilized for sheathing and finish.

6. Some disadvantages of joist and rafter construction are (one wrong): a. it takes a long time; b. the building is exposed to the weather longer; c. builder must utilize scaffolding; d. it requires special sheathing.

7. In joist and rafter roof framing, the rafters are put in place after the _____ have been fastened in place.

8. If it weren't for the _____, the rafters would spread and push out the exterior walls.

9. Complete the following statements on laying out a common rafter:

   a. The top of the rafter rests against the ___Ridge___ board.

   b. The cut at the top of the rafter is called a top or ___Plumb___ cut.

   c. The bottom of the rafter rests on the ___Plate___.

   d. The cut on the bottom of the rafter is called the level or ___seat___ cut.

   e. Draw the ___Plumb___ line along the edge of the tongue.

10. Three ways of figuring the length of a common rafter are:

   a. By stepping off the length with a ___framing___ square.

   b. By using the ___Pathagorean___ theorem.

   c. By applying the unit length obtained from the ___rafter___ table on the framing square.

(Continued on next page)

_____ F _✓_

_____ b _✓_

a. _____plumb_____ _✓_

b. _____feet_____ _✓_

_____allowance_____ _✓_

_____birds mouth_____ _✓_

_____saddle_____ _✓_

_____T_____ _✓_

_____F_____ _✓_

11. The theoretical length of a rafter and the actual length are the same. (T or F)

12. In a house that is 28′ wide with a 1/2 pitch roof, the theoretical length of a common rafter is *(one right)*: a. 124″; b. 188″; c. 215″; d. 230″.

13. The step-off method for finding the theoretical rafter length is done as follows:

    a. Place the square on the rafter with the tongue on the ___plumb___ cut.

    b. Step off the cut of the roof on the rafter stock as many times as there are ___feet___ in the total run.

14. The length of a rafter as shown on the rafter table must be reduced by half the thickness of the ridge board. This is called a ridge _____.

15. When a rafter has an overhang, it has a notch called a _____.

16. The bird's-mouth and the plumb cut at the ridge can be eliminated from a rafter by using a _____ brace.

17. Two rafters should be cut and used to see how the heel cut and top cut fit before cutting remaining rafters. (T or F)

18. Rafters are usually leaned against the building with the ridge cut down. (T or F)

3/12 slope is 1/8 pitch

Name _____

Score: (34 possible) _____

Study Unit 33
Hip and Valley Rafters

_____ hip _____ ✓

1. The roof member that forms a raised area is called a _____ rafter.

_____ Valley _____ ✓

2. The roof member that forms a depression in the roof is called a _____ rafter.

_____ 2 _____ ✓

3. Rafters that form a raised area or depression in the roof should be _____ inches deeper than common rafters.

a. _____ hip _____ ✓
b. _____ Common _____ ✓
c. _____ plate _____ ✓
d. _____ ridge _____ ✓
e. _____ Valley raffer _____ ✓
f. _____ valley jack raffer _____ ✓
g. _____ tails _____ ✓
h. _____ hip jack _____ ✓

4. Give the names of the parts drawn in the hip roof in Fig. 33-1.

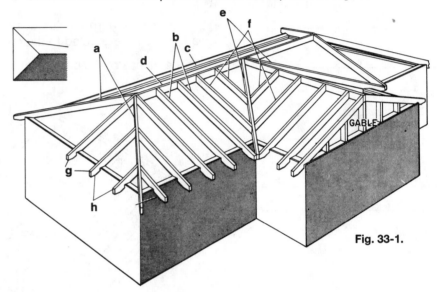

Fig. 33-1.

_____ right _____ ✓

5. The unit run of a hip rafter is the hypotenuse of a _____ triangle with the shorter sides each equal to the unit run of a common rafter.

_____ F _____ ✓

6. The unit length of a hip rafter may be found on the third line of the rafter table on the steel square. (T or F)

_____ bottom _____ ✓

7. The plumb and level lines on a hip or valley rafter are also referred to as the top and _____ cuts.

_____ ½ _____ ✓

8. If a hip rafter is framed against a ridge board, the shortening allowance is __1.06"__ the 45° thickness of the ridge piece.

_____ Angle _____ ✓

9. The hip rafter joins the ridge piece or the ridge end of the common rafter at an _____.

_____ F _____ ✓

10. The top line of the rafter table on the framing square is headed "Side Cut Hip or Valley Use." (T or F) *1st line is Common*

_____ F _____ ✓

11. The hip or valley rafter overhang is the same as the overhang for a common rafter on the same roof. (T or F)

(Continued on next page)

_____seat_____ ✓ 12. The bird's-mouth is formed by two cuts, namely, the heel cut and the _____ cut.

_____backing_____ ✓ 13. In adjusting the upper edges of a hip rafter so that it will not interfere with the application of sheathing, two methods can be used, namely, _____ and dropping.

_____T_____ ✓ 14. Dropping means to deepen the bird's-mouth to bring the top edge of the hip rafter down to the upper ends of the jacks. (T or F)

_____T_____ ✓ 15. Most roofs that contain valley rafters are equal-pitch roofs. (T or F)

_____T_____ ✓ 16. In framing an equal-span roof addition, the span of the addition is the same as the span of the main roof. (T or F)

_____valley_____ ✓ 17. An unequal-span roof addition can be framed with one long and one short _____ rafter.

_____T_____ ✓ 18. There are two methods of framing an unequal-span roof addition. (T or F)

_____T_____ ✓ 19. When framing a dormer without side walls, the upper edges of the headers must be beveled to the cut of the main roof. (T or F)

a. ___double common rafter___ 20. Give the names of the parts shown for a dormer in Fig. 33-2.

b. ___upper header___

c. ___main roof valley jack___

d. ___valley rafter___

e. ___lower header___

f. ___cripple common rafter___

g. ___dormer valley jack___

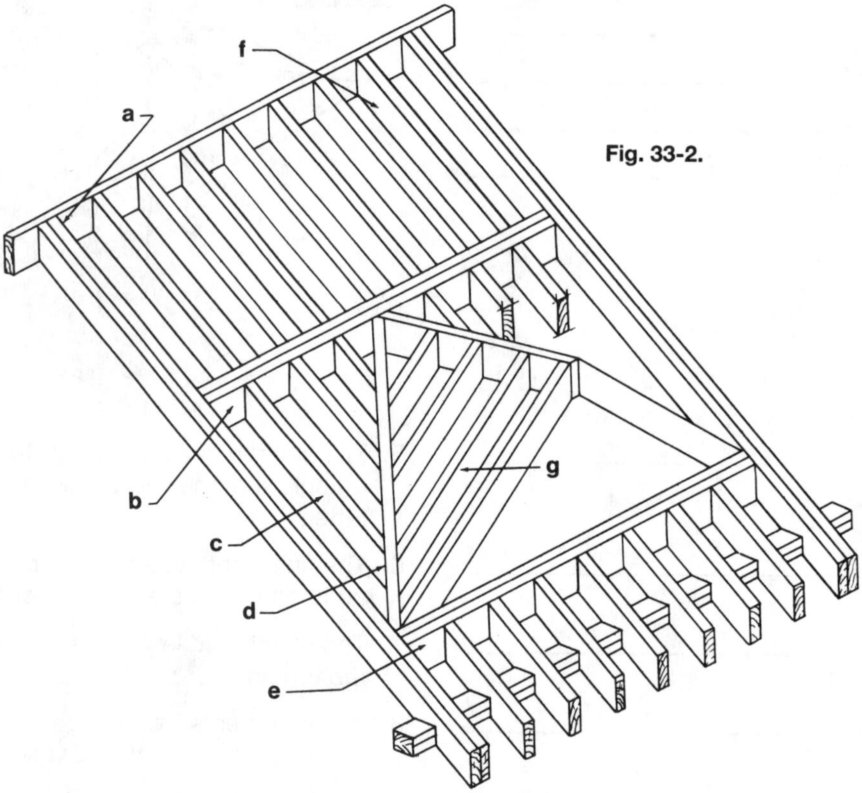

Fig. 33-2.

_____T_____ ✓ 21. A dormer may be constructed with side walls. (T or F)

82

Study Unit 34
Jack Rafters and Roof Framing

Shortened ✓	1. A jack rafter is a _____ common rafter.
b ✓	2. The following are the common kinds of jack rafters *(one wrong)*: a. hip jack; b. common jack; c. valley jack; d. cripple jack.
T ✓	3. When framing an equal-pitch roof, the unit rise of a jack rafter is the same as the unit rise of a common rafter. (T or F)
2 ✓	4. There are _____ types of cripple jack rafters.
F ✓ 16" oc	5. Jack rafters are usually spaced 12 inches apart. (T or F)
19.23 ✓	6. The length of the shortest hip jack when the jacks are spaced 16 inches on center and the unit rise is 8 inches is _____ inches.
T ✓	7. The best way to figure the total length of valley jacks and cripple jacks is to lay out a framing diagram. (T or F)
F ✓	8. In Fig. 36-4 in the text the run of valley jack No. 1 is the same as the run of hip jack No. 7. (T or F)
T ✓	9. In Fig. 36-4 in the text the run of valley jack No. 2 is the same as the run of hip jack No. 7. (T or F)
Twice ✓	10. In Fig. 36-4 in the text the run of valley cripple No. 14 is _____ the run of valley cripple No. 13.
F ✓	11. A hip jack rafter has a shortening allowance at the upper end consisting of one-half the 45° thickness of the hip rafter. (T or F)
T ✓	12. The bird's-mouth and overhang of a jack rafter are laid out the same as on a common rafter. (T or F)
c ✓	13. Before a building is ready for roof framing, the following must be completed *(one wrong)*: a. all framing is complete; b. all framing is plumbed and squared; c. exterior siding is applied; d. the ceiling joists are in place.
T ✓	14. On a gable roof the theoretical length of the ridge piece is equal to the length of the building. (T or F)
F ✓	15. On a gable roof, if there is an overhang, the ridge board is shorter than the length of the building. (T or F)
T ✓	16. A hip roof may have an equal or an unequal-span addition. (T or F)
F ✓	17. The ridge board will be the same length both in equal-span and in unequal-span additions. (T or F)
F ✓	18. The length of the ridge piece on a dormer is the same whether or not it has side walls. (T or F)
T ✓	19. The layout of rafter spacing can be found by checking the building plans or the roof framing plans. (T or F)

(Continued on next page)

_____ F _____ ✓ 20. For a gable roof the rafter locations are laid out on the ridge piece first. (T or F)

_____ T _____ ✓ 21. The rafters for a gable roof should butt directly opposite each other on the ridge board. (T or F)

_____ T _____ ✓ 22. There are two common methods of erecting the ridge board. (T or F)

_____ 4 _____ ✓ 23. Roof framing should be done from a scaffold with planking not less than _____ feet below the level of the main roof ridge board.

_____ T _____ ✓ 24. On a hip roof the ridge board and the common rafters extending from the ridge ends to the side walls are erected first. (T or F)

_____ F _____ ✓ 25. Common rafters on a hip roof must be plumbed. (T or F)

_____ 10D _____ ✓ 26. Hip rafters on a hip roof are toenailed to the plate corners with _____ nails, two on each side.

_____ F _____ ✓ 27. All hip jacks should be nailed on one side first and then all the jacks on the opposite side. (T or F)

_____ 3 _____ ✓ 28. Hip jacks are toenailed to hip rafters with 10d nails, using _____ for each jack.

_____ Collar _____ ✓ 29. Gable or double-pitch roof rafters are often reinforced by horizontal members called _____ beams.

_____ F _____ ✓ 30. The theoretical and actual lengths of the collar beams are identical. (T or F)

_____ T _____ ✓ 31. A gable roof may be framed with or without an overhang. (T or F)

_____ gambrel _____ ✓ 32. A gable roof with two slopes is called a _____ roof.

_____ T _____ ✓ 33. An advantage of the gambrel roof is that it provides additional space for rooms in the attic. (T or F)

_____ T _____ ✓ 34. A gambrel maximizes the roof area exposed to snow loads. (T or F)

_____ gable _____ ✓ 35. A shed roof is essentially one-half of a _____ roof.

_____ T _____ ✓ 36. A dormer may be framed into a shed roof. (T or F)

_____ F _____ ✓ 37. A flat roof is always perfectly flat. (T or F)

_____ beam _____ ✓ 38. A common kind of construction used with flat or low-sloped roofs is the post-and-_____ construction.

_____ T _____ ✓ 39. It is possible to combine the ceiling and roof elements into one system on a flat or low-sloped roof. (T or F)

_____ T _____ ✓ 40. Flat and low-pitched roofs usually require larger sized rafters than pitched roofs. (T or F)

_____ b _____ ✓ 41. Roof openings are commonly required for the following (one wrong): a. ventilator; b. overhang; c. chimney; d. skylight.

_____ T _____ ✓ 42. A chimney saddle can be fabricated on the ground. (T or F)

_____ T _____ ✓ 43. The number of rafters necessary for a building can be counted directly from a roof framing plan. (T or F)

_____ F _____ ✓ 44. It is not possible to estimate the number of rafters necessary for a house. (T or F)

Name _____

Score: (34 possible) _____

Study Unit 35
Roof Trusses

Page 85: Please strip into position.

_____ *T* _____ ✓

a. _____ *Triangular* ✓

b. _____ *gussets* ✓

c. _____ *support* ✓
_____ *triangle*

_____ *T* ✓
_____ *F* ✓

a. _____ *Walls* ✓

b. _____ *Partions* ✓

c. _____ *quickly* ✓
_____ *b* ✓

a. _____ *W*

b. _____ *King*

c. _____ *Scissor*

1. The roof truss can support loads over long spans. (T or F)

2. A simple truss is:

   a. Pieces put together to form a stiff framework of _____ shapes.

   b. Connected at the joints by plywood _____ in those cases where the truss is built on the job site.

   c. Able to hold loads over a long span without intermediate _____.

3. Gussets join together the parts of a _____ by means of nails, screws, bolts, or split-ring connectors.

4. Trusses usually come to the building site preassembled. (T or F)

5. The use of roof trusses adds considerably to material and labor costs. (T or F)

6. The use of trusses has the following advantages:

   a. No interior bearing _____ are needed.

   b. _____ can be placed anywhere in the home interior.

   c. Trusses can be put up very _____.

7. The types of trusses used for home construction include *(one wrong)*: a. scissors; b. Y-type; c. king-post; d. W-type.

8. Identify the trusses shown in Fig. 35-1.

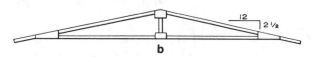

Fig. 35-1.

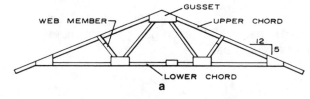

a. _____ *2*

b. _____ *1*

c. _____ *3*

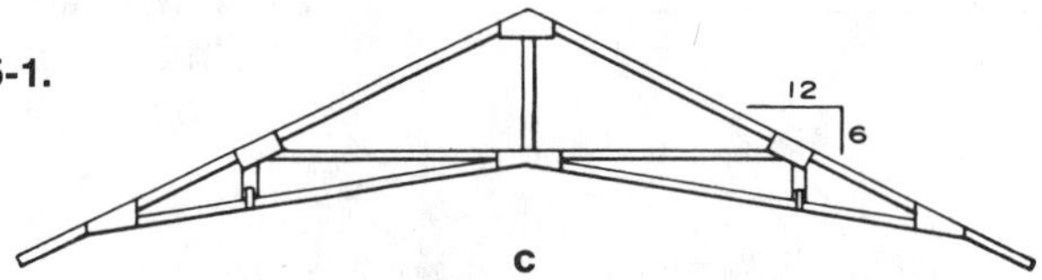

9. Match the items at the left to the descriptions at the right:

   a. scissors
   b. king-post
   c. W-type

   1. more economical for shorter spans because it has fewer pieces
   2. used for "cathedral" ceiling
   3. has more pieces with less distance between connections

(Continued on next page)

_____ 24" ___ ✓

_____ W ___ ✓

_____ 9 ___ ✓

_____ plywood ✓

_____ 19 ___ ✓

a _____ Manufactures ✓

b. _____ temperature ✓

c. _____ upright ✓

_____ staples ✓

_____ F ___ ✓

_____ F ___ ✓

_____ camber ✓

_____ weights ✓

_____ d ___ ✓

_____ b ___ ✓

_____ T ___ ✓

_____ 2 ___ ✓

10. The common spacing for roof trusses is _____ inches.

11. The truss used the most is the _____.

12. Factors most important in truss design are *(one wrong)*: a. kind of wood; b. snow and wind loads; c. slope of roof; d. weight of roof itself.

13. Gusset plates used on trussed rafters are made of metal or _____.

14. This is a math exercise. If it takes 3 workers 2 hours to install roof trusses in 2,000 square feet of ceiling area, it would take them _____ hours to install roof trusses in 19,000 square feet of ceiling area.

15. Follow these instructions when assembling a truss on the job:

    a. When mixing and using glue, follow the _____ directions.

    b. Do the gluing under controlled _____ conditions.

    c. Lift and store trusses in an _____ position.

16. Nails or _____ may be used to supply pressure until the glue is set when applying the gussets.

17. A truss can be put into position in the roof immediately after assembly. (T or F)

18. Building trusses on the job requires extensive equipment. (T or F)

19. When building a truss, don't forget to provide for a slight curvature, called _____.

20. The basic parts of a roof truss are usually connected at the joints by rectangular connector _____ made of metal.

21. A truss can be fastened to the wall plates *(one wrong)*: a. by toenailing; b. by a metal bracket; c. with a framing anchor; d. with adhesive.

22. The best method for fastening a truss to the wall plate is *(one right)*: a. by toenailing; b. by a metal bracket; c. with a framing anchor; d. with adhesive.

23. It is estimated that the use of roof trusses results in a materials savings of about 30% on roof members and ceiling joists. (T or F)

24. It will take three workers about _____ hours to install roof trusses on an average size home, with attached garage, containing about 2,000 square feet of ceiling area.

Name _____

Score: (41 possible) _____

Study Unit 36
Roof Sheathing

_____ T _____ ✓

_____ plywood _____ ✓

_____ laminated _____ ✓

_____ exposed _____ ✓

_____ F _____ ✓

_____ F _____ ✓

_____ F _____ ✓

_____ closed _____ ✓

_____ c _____ ✓

_____ 8D _____ ✓

_____ T _____ ✓

a. _____ base _____ ✓

b. _____ joints _____ ✓

c. _____ waste _____ ✓

d. _____ design _____ ✓

_____ T _____ ✓

_____ T _____ ✓

_____ F _____ ✓

_____ T _____ ✓

1. Roof sheathing is considered a structural part of a building. (T or F)

2. Lumber roof sheathing and _____ are two good sheathing choices for homes with pitched roofs.

3. Lumber or _____ roof decking is sometimes used in homes with exposed ceilings.

4. Manufactured fiber roof decking can be adapted to _____ ceiling applications.

5. The hardwoods are the wood species used for roof sheathing boards. (T or F)

6. Roof sheathing boards to be covered with asphalt shingles do not have to be seasoned. (T or F)

7. Board roof sheathing is always laid horizontally. (T or F)

8. When roof sheathing boards are laid with no spaces between, it is called _____ sheathing.

9. Roof boards should give solid and continuous support for *(one wrong)*: a. roofs where snow and wind-driven conditions prevail; b. roofs made of asphalt shingles; c. roofs made of wood shakes; d. roofs made in cold climates.

10. Lumber roof sheathing is nailed to each rafter with two _____ nails.

11. Boards should be laid in such a way that they bear on at least two rafters. (T or F)

12. As a sheathing material, plywood:

    a. Provides a smooth, solid _____.

    b. Provides a minimum number of _____.

    c. Costs less by cutting _____ to a minimum.

    d. Offers flexibility in _____.

13. Plywood sheathing is very durable. (T or F)

14. One disadvantage of plywood sheathing is the considerable time it takes to install it. (T or F)

15. Face grain of plywood sheathing should run in the same direction as the rafters. (T or F)

16. If plywood roof sheathing is not the exterior type, it should not be exposed to the weather. (T or F)

(Continued on next page)

a. _____ 2 _____ ✓

b. _____ 4 _____ ✓

c. _____ 3 _____ ✓

d. _____ 1 _____ ✓

_____ platform _____ ✓

a. _____ base _____ ✓

b. _____ ceiling _____ ✓

c. _____ permanent _____ ✓

d. _____ floor _____ ✓

e. _____ grades _____ ✓

_____ Select _____ ✓

_____ Commerical _____ ✓

_____ T _____ ✓

_____ 12 _____ ✓

_____ thickness _____ ✓

_____ a _____ ✓

_____ twice _____ ✓

_____ Galvenized _____ ✓

_____ splines _____ ✓

_____ strength _____ ✓

_____ d _____ ✓

_____ 3/4 _____ ✓

17. Match the plywood thickness at the left to the roofs for which they should be used at the right:

a. 3/8″
b. 1/2″
c. 5/16″
d. 5/8″

1. slate or tile and rafter spacing of 24″
2. asphalt shingles and rafter spacing of 24″
3. asphalt shingles and rafter spacing of 16″
4. slate or tile and rafter spacing of 16″

18. To help the worker apply plywood roof sheathing, a roof _____ may be constructed.

19. Roof decking has the following good points:

a. Is an excellent _____ for any roofing material.

b. Makes a ready-to-finish interior _____.

c. Provides a solid, _____ roof deck.

d. Is strong enough for use as _____ decking.

e. Comes in many _____, patterns, and sizes.

20. _____ grade decking is the best choice when appearance is important.

21. The grade of decking to choose when appearance and strength are not of prime importance is _____ grade.

22. Decking can be purchased with tongue and groove edges. (T or F)

23. Decking comes in nominal widths of 4 to _____ inches.

24. Check the manufacturer's recommendations to get the correct _____ for the span for all types of decking.

25. Decking should be installed (one wrong): a. with all the end joints in line; b. with each plank bearing on at least one support; c. with a 2° angle cut in the butting pieces; d. in a controlled random laying pattern for best economy.

26. Decking should be fastened with common nails which are _____ as long as the thickness of the nominal plank.

27. _____ common nails have better holding power than bright common nails.

28. Metal _____ are recommended on decking end joints of 3″ and 4″ material for better alignment, appearance, and strength.

29. Roof decking made of wood fiber combines the advantages of _____ and insulation.

30. Wood fiber roof decking (one wrong): a. is weatherproof; b. comes coated in a wide variety of colors; c. is ideal for shingles on all types of buildings; d. is fairly expensive.

31. For a roof with a chimney, the sheathing should clear the brick by _____ inch.

Study Unit 37
Roof Coverings

_____ b ✓

_____ T ✓

a. _____ 4 ✓

b. _____ 2 ✓

c. _____ 6 ✓

d. _____ 3 ✓

e. _____ 1 ✓

f. _____ 5 ✓

g. _____ 7 ✓

1. Roofing materials for pitched roofs include *(one wrong)*: a. shingles; b. hot asphalt or tar; c. slate or tile; d. galvanized metal.

2. If a roof is shingled, the shingles should always overlap. (T or F)

3. Match the items at the left to the descriptions at the right:
   a. sidelap or endlap
   b. coverage
   c. toplap
   d. square
   e. shingle butt
   f. exposure
   g. headlap

   1. lower exposed edge of a shingle
   2. amount of protection overlapping shingles provide
   3. amount of roofing needed to cover 100 sq. ft.
   4. shortest distance in inches two shingles overlap each other horizontally
   5. shortest distance in inches between exposed edges of overlapping shingles
   6. the width of a shingle minus its exposure
   7. shortest distance in inches from lower edges of overlapping shingle to upper edge of unit in second course below

_____ F ✓

_____ d ✓

_____ Slope ✓

(⅙ pitch) _____ 4/12 ✓

_____ c ✓

_____ b ✓

_____ F ✓

_____ d ✓

_____ low ✓

4. Slope and pitch are the same. (T or F)

5. Many roofing accessories are needed including *(one wrong)*: a. underlayment; b. roofing cements; c. roofing nails; d. fascia.

6. The amount of overlap of shingles depends largely on the kind of shingle and the _____ of the roof.

7. A roof that rises 4″ for each 12″ of run is designated as a _____ slope.

8. A roof shingled with slate or asphalt should have a slope of at least *(one right)*: a. 2 in 12; b. 3 in 12; c. 4 in 12; d. 6 in 12.

9. A built-up roof should slope no more than *(one right)*: a. 2 in 12; b. 3 in 12; c. 4 in 12; d. 6 in 12.

10. All types of shingles require application of underlayment. (T or F)

11. A good underlayment is *(one right)*: a. waterproof paper; b. coated felt; c. laminated paper; d. asphalt-saturated felt.

12. It is important to choose an underlayment with a _____ vapor barrier.

(Continued on next page)

a. _____Wood_____ ✓

b. _____moisture_____ ✓

c. _____Rain_____ ✓

_____sheathing_____ ✓

_____flashing_____ ✓

a. _____Ventilator_____ ✓

b. _____Chimney_____ ✓

c. _____Wall_____ ✓

d. _____Valley_____ ✓

_____F_____ ✓

_____Snap_____ ✓

_____C_____ ✓

_____T_____ ✓

_____rake_____ ✓

_____T_____ ✓

_____C_____ ✓

_____Cutouts_____ ✓

a. _____Center_____ ✓

b. _____hip_____ ✓

c. _____5"_____ ✓

_____F_____ ✓

_____T_____ ✓

13. Underlayment serves the following purposes:

   a. Keeps asphalt shingles from direct contact with _____ sheathing, which could cause a chemical reaction.

   b. Protects roof sheathing from absorbing _____ before shingles are applied.

   c. Prevents _____ from driving below shingles onto roof sheathing.

14. Underlayment should be applied immediately after _____.

15. A metal or other special material applied to ward off water seepage is called _____.

16. Some areas that need material to prevent water seepage are:

   a. At the intersection between roof and soil stack or _____.

   b. Around _____.

   c. At the point where a _____ intersects a roof.

   d. In the _____ of a roof.

17. Homes in all climates require application of eaves flashing. (T or F)

18. The edges of a roof are protected from leaks by the installation of _____ edges.

19. Roof shingles made of asphalt come in the following basic kinds (one wrong): a. individual shingles of a very large size; b. individual shingles, either interlocking or staple-down; c. asphalt and wood combination; d. strip shingles.

20. If a roof is longer than 30′, asphalt shingles should be applied starting at the center. (T or F)

21. On a roof shorter than 30′, asphalt shingles can be applied from either _____.

22. The pattern of laying asphalt strips can be varied with full or cut strips. (T or F)

23. Square-butt strip shingles are laid in variations, as follows (one wrong): a. cutouts breaking joints on halves; b. cutouts breaking joints on thirds; c. cutouts breaking joints on fourths; d. random spacing.

24. When laying each course of shingles, the lower edges of the butts should be aligned with the top of the _____ on the underlying course. 5″

25. Techniques to follow for applying asphalt shingles to hips and ridges are:

   a. Bend each shingle lengthwise down the _____ with an equal amount on each side of the hip or ridge.

   b. Begin at a _____ or at one end of a ridge.

   c. Use a _____ inch exposure.

26. The closed method of application is always used for valleys. (T or F)

27. A roof with a 3-in-12 slope is considered low pitch. (T or F)

a. _____ double _____

b. _____ eves _____

c. _____ adhesives _____

_____ interlocking _____

_____ shaker _____

_____ a _____

_____ d _____

_____ T _____

_____ d _____

a. _____ wood _____

b. _____ felt _____

c. _____ tar _____

d. _____ asphalt _____

e. _____ gravel _____

_____ corragated _____

_____ F _____

_____ 1" _____

28. For a low-pitch roof, strip shingles should be applied with the following requirements:

 a. Underlayment of _____ thickness.

 b. Flashing strip on the _____ cemented in place.

 c. _____ applied to shingles at the factory.

29. Lock-down, also called _____, shingles should be chosen for areas subject to high winds.

30. Hand-cut wood shingles are called _____.

31. Hand-cut wood shingles come in lengths of (one wrong): a. 12"; b. 18"; c. 24"; d. 32".

32. Regular wood shingles come in lengths of (one wrong): a. 16"; b. 18"; c. 24"; d. 32".

33. Wood shingles should extend out over the eave or rake. (T or F)

34. Characteristics of roll roofing are (one wrong): a. can be applied over old roofing; b. is less attractive in appearance than other roofing materials; c. is possibly less durable than asphalt shingles; d. is quite costly.

35. A typical built-up roof consists of:

 a. Decking made of _____.

 b. Layers of roofer's _____.

 c. Each layer mopped down with asphalt or _____.

 d. Final coating of tar or _____.

 e. Topping of roofing _____ or roll roofing.

36. An ideal roofing for garages and sheds is _____ metal roofing.

37. Gutters must be made of metal. (T or F)

38. Metal gutters should be installed so they drain toward the downspouts at a slope of _____ inch in 10 feet.

Study Unit 38
Ventilation

_____

1. If the ceiling area equals 1,200 square feet, the net area for the louvered gate opening should be *(one right)*: a. 2 square feet; b. 3 square feet; c. 4 square feet; d. 5 square feet.

_____

2. Louvered openings in the end walls of gable roofs should be as close to the ridge as possible. (T or F)

_____

3. A soil cover should be a _____ barrier such as plastic film, roll roofing, or asphalt-laminated paper.

_____

4. When there are rooms in the attic with sloping ceilings, the insulation should follow the roof slope and be so placed that there is a free opening between boards of at least *(one right)*: a. 1/2"; b. 1 1/2"; c. 1 3/4"; d. 2 1/4".

5. Refer to Fig. 38-1, which shows a variety of outlet ventilators. Match each ventilator with its description.

a. _____   a. triangular

b. _____   b. square

c. _____   c. half-circle

d. _____   d. vertical

e. _____   e. soffit

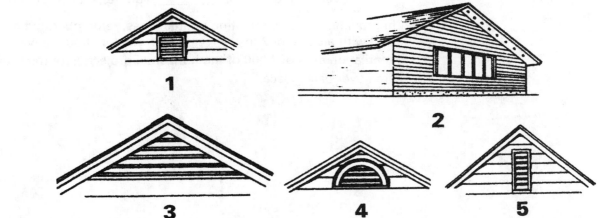

Fig. 38-1

_____

6. Condensation problems are greatly reduced by good ventilation. (T or F)

_____

7. A greater ratio of ventilation area is required in some types of flat roofs than in pitched roofs. (T or F)

_____

8. The size of the vent area in a home is relatively unimportant. (T or F)

(Continued on next page)

9. In ventilating the crawl space beneath a house:

a. _____

a. The soil cover should be a vapor barrier such as _____ film, roll roofing, or asphalt-laminated paper.

b. _____

b. Provide at least _____ foundation wall vents near corners of the building if there is no basement area.

c. _____

c. Wall vents are not required if there is a partial _____ open to a crawl space and if there is some type of operable window.

d. _____

d. The total free (net) area of the ventilators should be equal to 1/160 of the _____ area when no soil cover is used.

e. _____

e. The use of a ground _____ is recommended under all conditions.

10. Movement of air from inside to outside of a house can be provided by:

a. _____

a. _____ with screens in the gable ends of an attic.

b. _____

b. A slot under eaves and a sheet-metal _____ near the peak in a house with a hip roof.

11. In ventilating a house:

a. _____

a. If the ceiling area is 5400 square feet, the total net area of the ventilators in the gable roofs is _____ square feet.

b. _____

b. In ventilating a crawl space with a ground area of 1280 feet, a total net ventilating area of _____ square feet is required.

c. _____

c. For a crawl space with a ground area of 960 square feet, a total net ventilating area of _____ square feet is required.

d. _____

d. For a crawl space with a ground area of 2880 square feet, a total net ventilating area of _____ square feet is required.

e. _____

e. In providing cross-ventilation in attic areas, remember that a roof with a slope of 2 in _____ or greater should have a ventilation area of 1/300 of the horizontal projection of the roof area over each space.

Name _____

Score: (26 possible) _____

Study Unit 39
Roof Trim

1. Match the terms at the left with the descriptions at the right:

a. _____

b. _____

c. _____

    a. cornice
    b. eave
    c. rake

    1. the edge at the end of a gable roof
    2. the rafter-end overhang of a roof
    3. finish on the outside just below the eave

_____

2. A gable roof has eaves on all four ends. (T or F)

_____

3. The three general types of cornice are close, open, and _____.

_____

4. A strip that fits close under the eave is called a _____.

a. _____

b. _____

c. _____

d. _____

5. Name the parts of the cornices in Fig. 39-1:

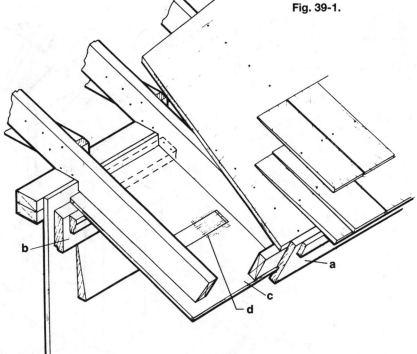

Fig. 39-1.

_____

6. A strip nailed to the tail plumb cuts of the rafters is called a _____.

_____

7. In a _____ cornice, the rafter overhang is entirely covered with roof covering, fascia, and bottom strip.

_____

8. The bottom strip is commonly referred to as a plancier, or _____.

(Continued on next page)

a. _____

b. _____

c. _____

9. Identify the cornices shown in Fig. 39-2:

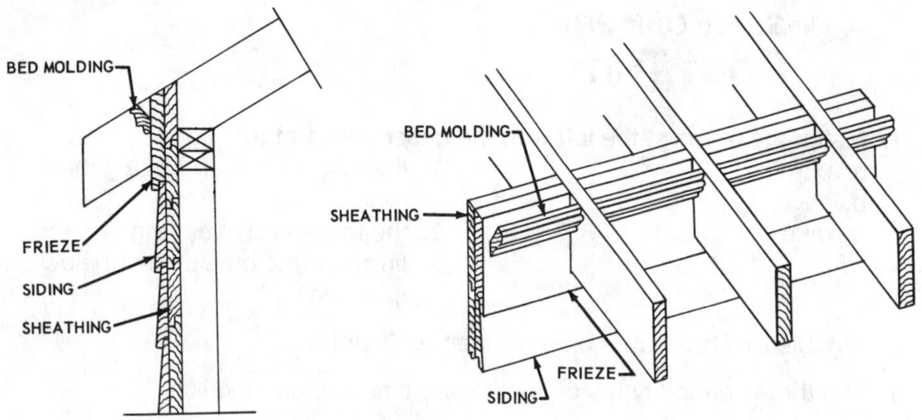

BED MOLDING

FRIEZE

SIDING

SHEATHING

BED MOLDING

SHEATHING

FRIEZE

SIDING

a

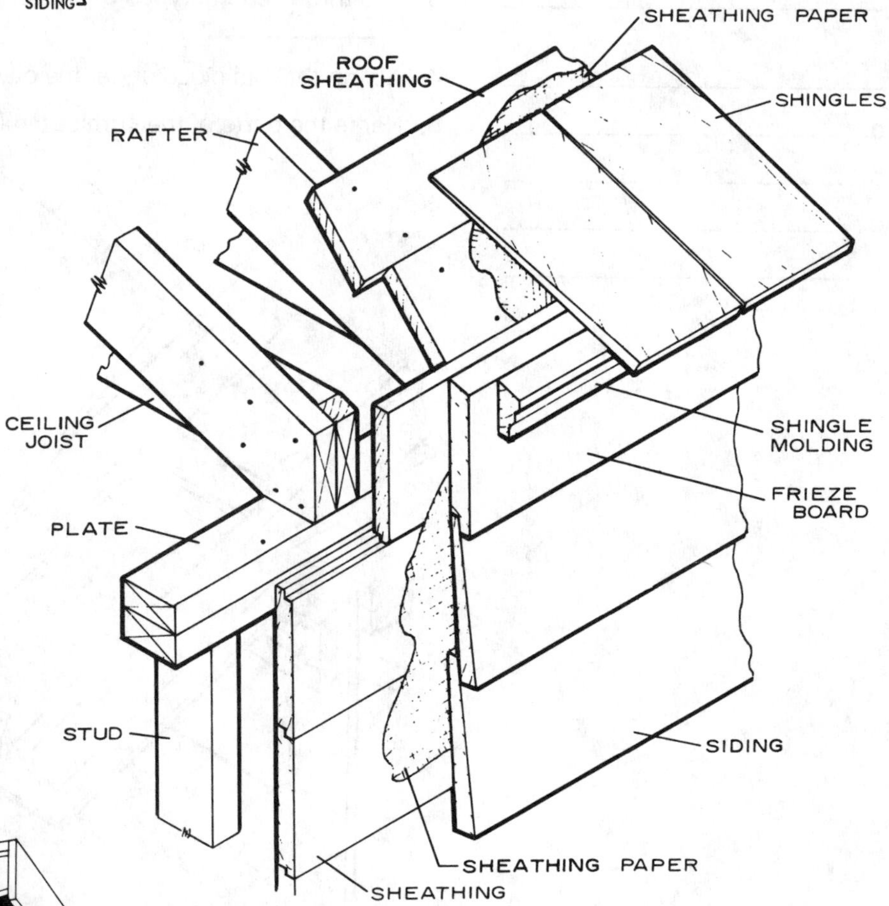

Fig. 39-2.

ROOF
SHEATHING

RAFTER

SHEATHING PAPER

SHINGLES

SHINGLE
MOLDING

FRIEZE
BOARD

CEILING
JOIST

b

PLATE

STUD

SIDING

SHEATHING PAPER

SHEATHING

c

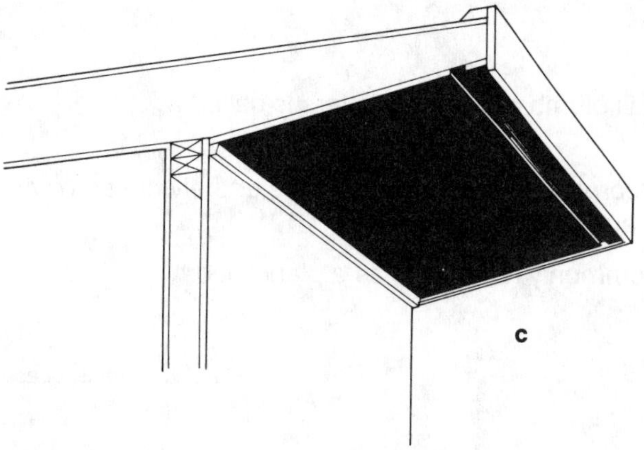

_____ 10. The quality of workmanship for an open cornice is unimportant. (T or F)

_____ 11. When constructing a box cornice, the rafter tails must be all in _____.

_____ 12. _____ is a popular material that presents a smooth, attractive surface for constructing a box cornice.

_____ 13. _____ board is a noncombustible material developed for use in soffits.

_____ 14. _____ panels are often used for soffits on the undersides of eaves.

_____ 15. A prefinished, low-maintenance, non-ferrous rust-free box cornice material is _____.

_____ 16. The ends of cornices on a gable roof must be finished. This finish is called the cornice _____.

17. Match the cornice returns at the left to the descriptions at the right:

a. _____
b. _____
c. _____

a. narrow cornice with box return
b. wide overhang at cornice and rake
c. narrow box cornice and close rake

1. used when there are wide overhangs at both sides and ends of the house
2. frieze board of the gable end joins the frieze board or fascia of the cornice
3. fascia board and shingle molding of the cornice are carried around the corner of the rake projection

To be used with Unit 43 | Name ____________________

Pages 499-530 | Score: (30 possible) ______

Study Unit 40

Windows and Skylights

a. ____________________

b. ____________________

c. ____________________

d. ____________________

e. ____________________

f. ____________________

1. Name the windows shown in Fig. 40-1.

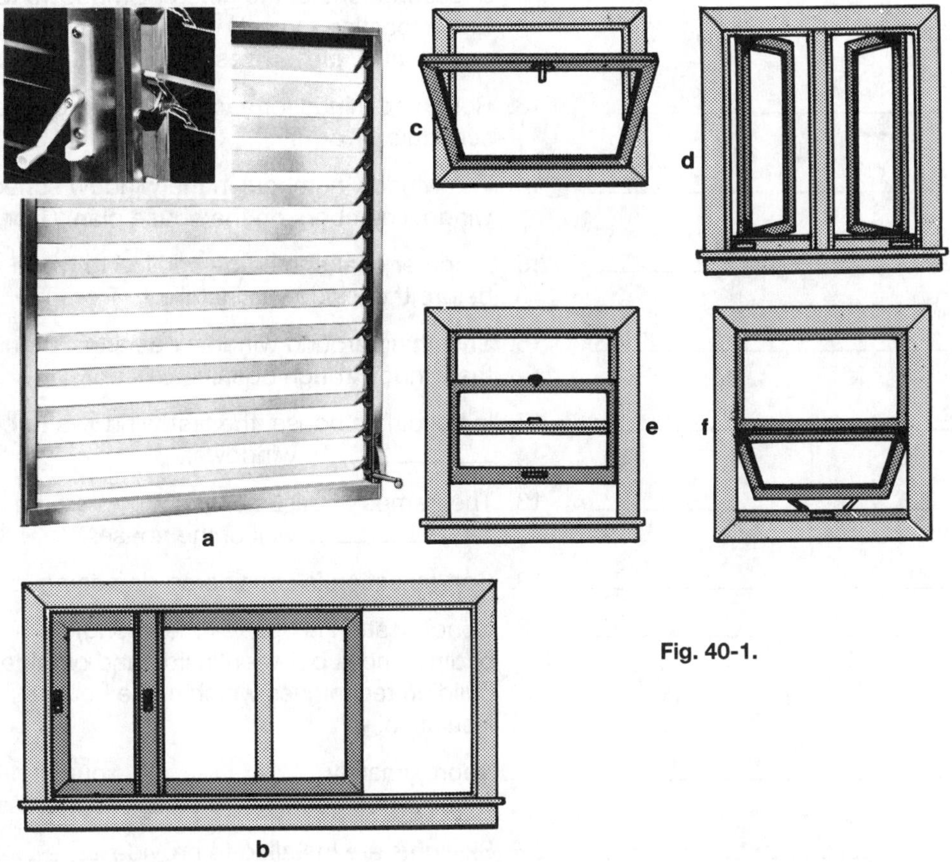

Fig. 40-1.

____________________ 2. Windows are manufactured items brought to the building site fully assembled. (T or F)

____________________ 3. Window frames are made of wood, aluminum, or ____________.

____________________ 4. The window best known and most widely used is the ____________ window.

____________________ 5. A window hinged at the sides that is designed to swing inward or outward by means of a crank is called the ____________ window.

____________________ 6. An advantage of the ____________ window is better ventilation because the entire window opens.

____________________ 7. A window designed solely for providing decoration and light is the ____________ window.

____________________ 8. The ____________ window is hinged at the top so the bottom can swing out.

(Continued on next page)

_____ 9. A window hinged at the bottom so the top can swing inward is called a _____ window.

_____ 10. Horizontal _____ windows can be used effectively to create a wall of windows.

_____ 11. _____ are often built into a roof to provide ventilation and light.

_____ 12. A window schedule contains the following information (one wrong): a. descriptions of the different windows to be installed in the house; b. the location of each window; c. the cost of the windows; d. sash openings; e. glass sizes.

_____ 13. Rough openings for each window are always given on the window schedule. (T or F)

_____ 14. The window numbers in the window schedule correspond with the window numbers on the house plan. (T or F)

_____ 15. A primer coat should be applied to wood window frame members before the window is installed. (T or F)

_____ 16. Sheathing around window openings should always be applied flush with the window rough openings. (T or F)

_____ 17. In building a house, the first windows to be installed are the _____ windows.

_____ 18. The frames of basement windows are an integral part of the _____ wall of the house.

_____ 19. Condensation is considered damaging to a home. (T or F)

_____ 20. Condensation is due to (one wrong): a. high humidity in the home; b. differences between inside and outside temperatures; c. improved building techniques which make houses more tightly built; d. poor insulation.

_____ 21. Approximately _____ hour(s) of labor will be needed for a window that contains 10 square feet or less of glass area.

_____ 22. Skylights are installed to provide _____ and light.

_____ 23. Glazing of skylights can be either glass or _____.

_____ 24. It is easier to install a skylight in an existing roof than in a new roof. (T or F)

_____ 25. All skylights can be opened for ventilation. (T or F)

Study Unit 41
Exterior Doors and Frames

a. _____

b. _____

c. _____

d. _____

e. _____

1. Identify the doors shown in Fig. 41-1.

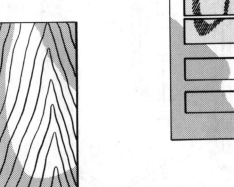

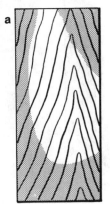

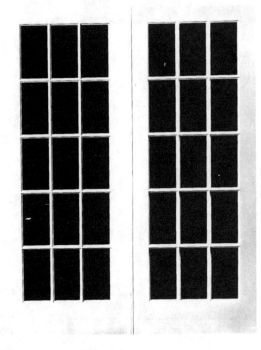

Fig. 41-1.

_____

a. _____

b. _____

c. _____

d. _____

e. _____

_____

_____

2. Another name for a French door is a _____ door.

3. Match the doors at the left to the descriptions at the right:
   a. French
   b. flush
   c. panel
   d. sliding
   e. combination

   1. facing applied to a core
   2. panels including inserts of screen or storm
   3. set to ride in a track
   4. made of many panes
   5. made of thin members which fit between the rails and stiles

4. Exterior door thickness is *(one right)*: a. 1 1/2"; b. 1"; c. 1 3/4"; d. 2".

5. Door jambs are always made of metal. (T or F)

(Continued on next page)

a. _____

b. _____

c. _____

d. _____

e. _____

f. _____

6. Name the parts of the doorframe shown in Fig. 41-2.

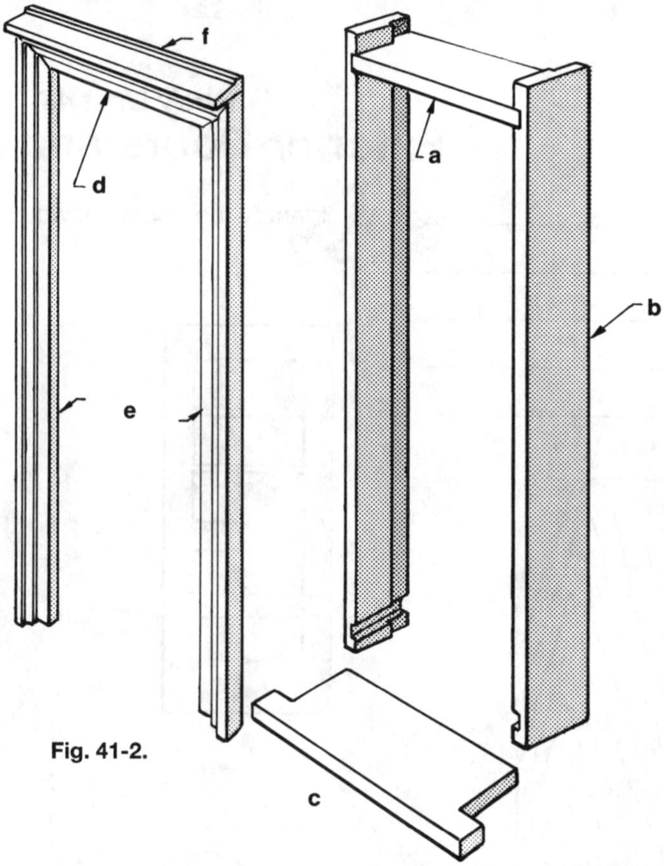

Fig. 41-2.

_____

_____

_____

a. _____

b. _____

c. _____

d. _____

e. _____

f. _____

_____

_____

_____

7. The front or main entrance should be at least _____ feet wide.

8. The rough opening for the doorframe should measure the same size as the frame. (T or F)

9. The doorframe should fit into the rough opening with the sill 2″ higher than the surface of the finished floor. (T or F)

10. Doors are expensive and should be cared for properly, as follows:

   a. Store under _____ away from moisture.

   b. _____ the top and bottom edges.

   c. Move a door by _____ it.

   d. Wear _____ when handling doors.

   e. Condition to the average _____ content of the locality.

   f. Store on _____ on a level surface to avoid warp or bow.

11. Doors should be finish-painted at the same time as the woodwork. (T or F)

12. A door should be fitted into the frame with a little leeway allowed for clearance. (T or F)

13. The edge of the door on which the _____ will go should be beveled.

_____ 14. A device used for holding a door upright for planing and hardware installation is called a door _____.

_____ 15. The number of hinges usually installed on an exterior door is _____.

16. Complete the following descriptions of garage doors:

a. _____ a. The door that opens outward and is held by door holders is the _____ door.

b. _____ b. The door hung from a track above is called a sliding _____ door.

c. _____ c. The _____ door pivots on a track in the ceiling and has rollers at the center and top of the door.

d. _____ d. A door with rollers at each section fitted into a track at the side of the door and ceiling is the _____ overhead door.

_____ 17. The garage door best protected from weather is the _____ door.

Study Unit 42
Exterior Walls: Siding and Brick Veneer

a. _____

b. _____

c. _____

_____

_____

_____

_____

_____

_____

_____

_____

1. The variety of exterior wall coverings available is due to:

   a. The many different kinds of _____ used.

   b. The many different shapes and _____.

   c. The surface _____.

2. Siding made of wood is about the most common of all exterior wall coverings. (T or F)

3. Wood siding should *(one wrong)*: a. be easy to work; b. be made of hardwood; c. hold paint well; d. be free from warp.

4. The most common types of wood siding are *(one wrong)*: a. shakes; b. board; c. bevel; d. drop.

5. Wood siding that is nailed directly to the studs and that acts as both wall covering and sheathing is called _____ siding.

6. Wood siding should be treated with water repellent before installation. (T or F)

7. Bevel siding is always applied horizontally. (T or F)

8. The amount of overlap of rows of bevel siding is the same for all widths. (T or F)

9. A simple device that can be made to ensure accuracy and increase efficiency in installing vertical siding is called a truss pole. (T or F)

10. When marking pieces of siding which will fit into spaces such as between two windows, a siding gauge called a _____ is very useful.

a. _____

b. _____

c. _____

d. _____

e. _____

f. _____

11. Wood siding nails should:

   a. Be strong enough so _____ is not necessary.

   b. Not discolor or _____ the siding.

   c. Be easy to _____ into the siding.

   d. Not cause _____, even at the edge.

   e. Not pop out after being installed _____ with the siding.

   f. Be _____ resistant.

a. _____

b. _____

12. The best wood siding nails are:

   a. High tensile strength _____ nails.

   b. _____ nails.

(Continued on next page)

a. _____

b. _____

c. _____

d. _____

e. _____

f. _____

_____

a. _____

b. _____

c. _____

_____

_____

_____

_____

_____

_____

_____

_____

_____

a. _____

b. _____

c. _____

_____

13. Moisture will create problems with wood siding unless:

   a. Joints in siding are sealed with _____.

   b. Siding pieces are cut carefully to insure proper _____.

   c. There is enough _____ between courses.

   d. Gutter joints and _____ are well fitted.

   e. _____ is installed in places where horizontal surfaces meet siding.

   f. The lowest edge of the siding is at least _____ inches above the ground.

14. The plywood best suited for siding is called medium-density _____ plywood.

15. Plywood makes a good siding because:

   a. It is more _____ and stronger than lumber sheathing.

   b. It can be applied vertically or _____.

   c. It cuts the _____ required for installation.

16. Hardboard siding can be described as follows (one wrong): a. tough; b. does not split or splinter; c. does not hold a finish as long as lumber; d. resistant to dents.

17. Another siding material that is applied the same as hardboard is _____ board.

18. Sheathing is required when hardboard is installed. (T or F)

19. Wood shingles can be purchased (one wrong): a. in several grades; b. in standard widths; c. in several lengths; d. either finished or untreated.

20. Wood shingles are applied to exterior walls by single- or double-coursing. (T or F)

21. When wood shingles are applied over _____ or fiberboard, horizontal nailing strips must first be nailed to the studs.

22. It is recommended that wood shingles be no wider than _____ inches.

23. Wood shakes are made by _____ different methods.

24. _____ shakes are made mostly by hand.

25. Wood shingles and shakes can be allowed to weather. (T or F)

26. Siding and shingles made of heat-resistant cement material must be applied:

   a. Over _____ sheathing.

   b. Over _____ paper that has been applied over the sheathing.

   c. In the same way as horizontal _____ siding.

27. Stucco is a good finish for exterior walls. (T or F)

_____

28. Stucco should not be used for a two-story house unless the house was built by the _____ framing method.

29. When stucco is used as an exterior finish on a house:

a. _____

a. The first coat should be forced through the _____.

b. _____

b. It should be applied in _____ coats.

c. _____

c. The temperature should be above _____ degrees F.

_____

30. A house covered with stone or brick veneer requires no sheathing. (T or F)

_____

31. Metal and plastic sidings have the following advantages (one wrong): a. come with a baked-on finish; b. require little care and maintenance; c. can be worked with standard tools; d. require no sheathing.

_____

32. In applying metal or plastic siding, nails should be driven very tightly. (T or F)

33. Metal or plastic siding should be applied:

a. _____

a. With an allowance for expansion and _____.

b. _____

b. With overlaps in the siding away from areas of _____.

_____

34. Brick veneer is relatively low in initial cost. (T or F)

_____

35. The following are common mason's tools (one wrong): a. trowel; b. level; c. hammer; d. shears.

_____

36. Common types of bricks include (one wrong): a. building; b. facing; c. back; d. fire.

_____

37. All brick can be classified as either modular or nonmodular. (T or F)

_____

38. Type-O mortar is recommended for nonload-bearing walls. (T or F)

_____

39. The standard joint thicknesses are (one wrong): a. 1/8″; b. 1/4″; c. 3/8″; d. 1/2″.

_____

40. A brick veneer wall must be supported on foundation. (T or F)

_____

41. Weep holes are placed in veneer walls to allow hot air to escape. (T or F)

_____

42. When new mortar is placed in an existing wall, it is called _____.

Study Unit 43
Thermal Insulation, Radiant Barriers, and Vapor Barriers

_____   1. Insulation used as part of the construction of a home (*one wrong*):
a. keeps the home warmer in winter; b. keeps the home cooler in summer; c. increases the amount of duct work needed; d. reduces the size furnace required.

_____   2. A material having insulation properties controls energy in the form of heat, sound, and _____.

3. Match the insulation materials at the left to the descriptions at the right:

a. _____   a. reflective   1. applied by spraying or in board form
b. _____   b. corrugated paper
c. _____   c. flexible   2. retards heat by means of radiation
d. _____   d. rigid
e. _____   e. loose fill   3. vegetable product made into sheets, squares, or boards
f. _____   f. urethane or polystyrene
g. _____   g. sprayed or foamed   4. blanket insulation made by spraying multiple layers of paper
                                                  5. made in the form of blankets or batts
                                                  6. inorganic fibrous material blown against clean surfaces
                                                  7. poured, packed, or blown into such areas of home as attic and sidewalls

_____   4. Most flexible insulation is manufactured with a paper or sheet material covering. (T or F)

_____   5. Reflective insulation is better for cold climates than for warm. (T or F)

_____   6. Such materials as wallboard, roof decking, and building boards are considered examples of _____ insulation.

_____   7. A unit used to measure the transmission of _____ is called a Btu.

_____   8. A Btu is the amount of heat which will raise the temperature of one pound of water two degrees Fahrenheit. (T or F)

                           9. Heating engineers and thermal experts have exact knowledge concerning the insulating qualities of recommended materials. (T or F)

_____   10. Heat loss from a house during cold weather can be reduced by insulating all walls, ceilings, roofs, and floors that separate heated from _____ spaces.

_____   11. Insulation for a house with an unheated crawl space should be placed between the floor joists or around the wall _____.

(Continued on next page)

_____

12. A house with an unheated attic should have insulation installed around the stairway and in the first-floor _____.

_____

13. A wall next to an unheated garage or _____ should be insulated.

_____

14. A house with air conditioning has the same insulation requirements as a house built in a cold climate. (T or F)

_____

15. Flexible insulation should be placed between framing members so that the tabs lap the edge of the studs and also the top and bottom _____.

_____

16. The fastening device usually used for installing flexible insulation is a hand _____.

_____

17. Reflective insulation should be installed tightly with no allowance for air space. (T or F)

_____

18. Radiant barriers are not effective in hot climates. (T or F)

_____

19. Radiant barriers must have a reflective material on both sides of the substrate. (T or F)

_____

20. One-sided radiant barriers must have the reflective surface facing the open air space in the attic. (T or F)

_____

21. The simplest method of installing a radiant barrier is on the attic floor. (T or F)

_____

22. All material emits energy by _____ radiation.

_____

23. The number 1 indicates the highest level of reflectivity. (T or F)

_____

24. A radiant barrier can be attached near the roof of a house. (T or F)

_____

25. Condensation of moisture can cause a problem when a radiant barrier is installed on the floor. (T or F)

_____

26. The fire rating of radiant barriers must be rated class _____ by the National Fire Protection Association.

_____

27. Vapor barrier should be put in place before application of loose fill insulation. (T or F)

28. A process of vapor barrier installation called "enveloping" entails the following:

a. _____

   a. It seals entire walls with rolls of _____ film.

b. _____

   b. It is applied over insulation with no vapor _____.

c. _____

   c. It should be fitted tightly around openings and _____ if necessary.

d. _____

   d. It should be applied over studs, plates, and window and door _____.

_____

29. Application of vapor barrier to a house insures complete freedom from vapor leakage. (T or F)

Name _____

Score: (21 possible) _____

Study Unit 44
Sound Insulation

_____

_____

_____

a. _____

b. _____

c. _____

_____

_____

_____

_____

_____

_____

_____

_____

_____

_____

1. Sound insulation has always been important in hotels. (T or F)

2. The following appliances increase the noise level in a home *(one wrong)*: a. television; b. radio; c. hot water heater; d. stereo.

3. A family room is sometimes known as an _____ living room.

4. Complete the following description of the way sound travels:

   a. Noises create sound _____.

   b. These radiate outward through the _____.

   c. They strike surfaces which _____.

5. The construction of a home has a high degree of influence on sound transmission within the home. (T or F)

6. Parts of a house such as a ceiling, wall, or floor are rated for their sound resistance by STC. These initials stand for Sound _____ Class.

7. The higher the STC number, the better the insulation against sound. (T or F)

8. Sound travels readily through *(one wrong)*: a. masonry; b. air; c. wall studs; d. electrical switches.

9. The floor with the highest INR rating is *(one right)*: a. 0; b. –2; c. –5; d. +2.

10. Two excellent examples of sound-absorbing materials are _____ tile and panels.

11. Normal speech can be understood quite easily in a room with a sound transmission class of _____.

12. A city has an average background noise level of _____ db.

13. The STC rating of 8″ concrete block is _____.

14. Sound that goes around poorly fitted doors follows what is called a _____ path.

15. A 1″ square hole in a wall rated at STC 50 can reduce the wall's performance to STC _____.

16. The sound insulation of double wall in detail D in Fig. 47-6 has an STC rating of _____.

(Continued on next page)

17. The STC rating of floor-ceiling combination of detail A in Fig. 47-7 is
_____.

18. A sound insulation product can be placed beneath a floor surface to reduce the transmission of _____ sounds.

19. The STC rating of floor-ceiling combination shown in detail B of Fig. 47-8 is _____.

_____

_____

_____

Study Unit 45
Plaster and Drywall

_____

_____

_____

_____

_____

_____

_____

_____

a. _____

b. _____

c. _____

d. _____

e. _____

_____

_____

_____

_____

_____

_____

1. Some good interior finish materials for walls and ceilings are *(one wrong)*: a. drywall; b. gypsum; c. pegboard; d. plaster.

2. The base for a plaster finish is called _____.

3. The types of plaster base used the most are *(one wrong)*: a. sheet metal; b. gypsum; c. wood stripping; d. insulating fiberboard.

4. As plaster dries, _____ can develop around openings and corners.

5. Plaster must be reinforced around openings and corners by the application of _____ metal lath.

6. Pieces of wood located around openings to act as guides in a wall to be plastered are called plaster _____.

7. Plaster is made of the following *(one wrong)*: a. plastic; b. sand; c. water; d. lime.

8. Plaster is usually applied in _____ coats.

9. Match the coatings at the left to the descriptions at the right:
   a. sand-float
   b. brown or leveling
   c. putty
   d. scratch
   e. double-up

   1. a finish coat containing lime and sand
   2. the first plaster coat
   3. a smooth finish coat containing no sand
   4. a coat which combines both scratch and brown coats
   5. a second plaster coat which is leveled during application

10. A combination of scratch and brown plaster coats is used on _____ or insulating lath.

11. A finish plaster coat that takes a good gloss paint or enamel finish is _____ finish.

12. Such materials as fiberboard and plywood are called _____ wall.

13. Drywall is always applied vertically to the wall. (T or F)

14. Drywall is fastened to the studs with nails, screws, or _____.

15. One advantage of drywall is that it can be applied without regard for conditions of moisture in framing members of the house. (T or F)

16. Most drywall used as wall covering is _____ inch thick.

(Continued on next page)

_____ 17. When drywall is nailed to the studs, begin at the outside and work toward the center. (T or F)

_____ 18. When a nail head moves in the drywall after it is driven, it causes a nail _____.

_____ 19. The _____ nailing system helps prevent movement between drywall and the nail head.

_____ 20. It is best to cut drywall with a _____.

_____ 21. Joints between drywall should be smooth. This is done by applying joint _____ and perforated tape.

_____ 22. Joints must be sanded. (T or F)

_____ 23. Drywall cannot be applied to ceilings. (T or F)

_____ 24. Drywall is sheet material made up of gypsum filler faced with _____.

_____ 25. Drywall can be applied with (one wrong): a. nails; b. cement; c. screws; d. adhesives.

_____ 26. When the relative humidity is 50% and the temperature is 80 degrees F., the drying time for joint compound is _____ hours.

Study Unit 46
Wood Paneling

_____ 1. Paneling is popular for walls and ceilings because of the _____ and warmth of wood.

_____ 2. Wood paneling can be installed directly over studs. (T or F)

_____ 3. The basic types of paneling are *(one wrong)*: a. plywood; b. particleboard; c. hardboard; d. solid wood strips.

_____ 4. Paneling comes in thicknesses from 1/4" to _____".

_____ 5. Paneling should be delivered to the room in which it is to be installed _____ hours before installation.

_____ 6. The sheet size of paneling is generally 4' × 10'. (T or F)

_____ 7. Masonry walls should be waterproofed before the studding or furring is applied. (T or F)

_____ 8. Shims behind furring strips should be wood _____.

_____ 9. A plywood panel fastened from the floor partway up a wall is called _____.

_____ 10. A partial plywood panel cut to go partway up a wall is usually _____ inches long.

_____ 11. A full-size plywood panel is _____ inches long, and three panels can be cut from it.

_____ 12. Plywood panel can be purchased in a wide variety of styles and veneers. (T or F)

_____ 13. Plywood panel must be installed unfinished. (T or F)

_____ 14. When it is necessary to cut a hole in a piece of installed plywood such as an outlet box, do the cutting with a _____ saw.

_____ 15. The same techniques for applying plywood to walls can be followed for fiberboard and hardboard. (T or F)

_____ 16. Wood paneling to be used for wall covering *(one wrong)*: a. comes finished or unfinished; b. should be seasoned; c. should be 10" or wider; d. should have correct moisture content.

_____ 17. If wood paneling is installed horizontally *(one wrong)*: a. fewer pieces are needed; b. the process takes a shorter time; c. it makes the room appear longer; d. it makes the room appear higher.

_____ 18. The nail size for 1/4" thick plywood should be _____d.

(Continued on next page)

_____ 19. When cutting openings for electrical outlets, a _____ cut can be made with a saber saw to start the cut.

_____ 20. A room with a perimeter of 66′ to 68′ will require _____ panels.

_____ 21. The perimeter of a room is calculated by adding the _____ of the four walls.

Study Unit 47
Wood and Vinyl Flooring

_____

_____

_____

_____

_____

_____

_____

_____

_____

_____

_____

a. _____

b. _____

c. _____

d. _____

e. _____

_____

_____

1. A finish flooring of hardwood strips requires an underlayment. (T or F)

2. The most popular hardwood floor is *(one right)*: a. maple; b. birch; c. oak; d. pecan.

3. Hardwood strip flooring *(one wrong)*: a. is expensive; b. has beautiful grain; c. is different from any other floor of its kind; d. can be laid in a hexagon pattern.

4. Strips for hardwood flooring *(one wrong)*: a. come in different thicknesses; b. come in different widths; c. are usually tongued and grooved to fit together; d. can be purchased in one uniform grade.

5. The most popular thickness for wood strips is *(one right)*: a. 1"; b. 25/32"; c. 1/2"; d. 19/32".

6. The most popular width for wood strips is *(one right)*: a. 1 1/4"; b. 1"; c. 2"; d. 2 1/4".

7. Because flooring strips are made of hardwood, they come in random lengths. (T or F)

8. Grading of hardwood strip flooring is supervised by the Department of the Interior. (T or F)

9. There are _____ organizations which enforce grading practices.

10. Red oak is graded higher than white oak. (T or F)

11. Hardwood flooring is kiln-dried before delivery to the building site. (T or F)

12. Special care should be taken when storing hardwood flooring:

    a. The storage room should have _____ windows.

    b. The floor on which flooring is stored should be at least _____ inches above ground.

    c. The building should not be cold or _____.

    d. In winter the storage room should be heated to at least _____ degrees.

    e. Before installing, store flooring at the installation site at least four or five _____.

13. Flooring should be installed after the following construction operations have been completed *(one wrong)*: a. plastering; b. electrical wiring; c. interior trim; d. plumbing.

14. Flooring is ready for installation immediately after the bundles are unwrapped. (T or F)

(Continued on next page)

Carpentry and Building Construction Student Workbook
Protected by Copyright

117

_____

_____

_____

_____

_____

_____

15. Before installation begins, a thorough inspection of the _____ should be made.

16. The last step before actual installation is to cover the subfloor with asphalt-coated building _____.

17. Studying the floor _____ will help the carpenter decide where to start the finish floor installation.

18. When strip flooring will run from room to hall to another room, the installation should begin in the hall. (T or F)

19. When strip flooring is laid up to a room with a door and a different kind of flooring, the strip flooring should end under the center of the door when it is _____.

20. The best way to lay strip flooring is (one wrong): a. parallel to the length of the room; b. running continuously between adjoining rooms; c. with ends of strips aligned; d. crosswise of the room's length if room is sufficiently wide.

21. Nailing should be done as follows:

a _____

b. _____

c. _____

d. _____

    a. If flooring strips are 25/32″ thick and 1 1/2″ or more wide, use 7d or _____d nails.

    b. Either _____ nails or cut steel nails are satisfactory.

    c. If flooring nails chosen are steel-wire flooring nails, they should be _____d.

    d. Steel-wire flooring nails should be coated with _____.

22. Wood strip flooring cannot be laid in a house with a concrete slab. (T or F)

_____

_____

23. Strip flooring that has been surfaced at the factory requires no sanding after installation. (T or F)

24. Many traditional floor finishes have been replaced by _____ finishes.

_____

25. Safety instructions for applying floor finishes include plenty of ventilation and the use of a _____.

_____

26. Three types of urethane finishes include (one wrong): a. oil-modified urethanes; b. moisture-cured urethanes; c. color-based urethanes; d. water-based urethanes.

_____

27. Acid-curing finishes contain formaldehyde. (T or F)

_____

28. Hardwood floor wax is available in _____ or liquid form.

_____

29. The wood preferred for plank flooring is _____.

_____

30. Pattern floors are also known as _____ and design floors.

_____

31. Strip flooring that is 3/8″ × 2″ requires _____ board feet to cover an area of 100 square feet.

_____

32. The most common type of resilient flooring is made of _____.

_____

33. An underlayment of plywood should be at least _____″ thick.

34. Vinyl tile can be laid directly over wood flooring. (T or F)

_____

35. Asphalt tile that is 9″ × 9″ square will require _____ hours of labor if 100 square feet are to be installed.

Study Unit 48
Ceramic Tile

_____

_____

_____

_____

_____

_____

_____

_____

_____

_____

_____

_____

_____

_____

_____

_____

_____

_____

_____

_____

_____

_____

_____

_____

1. Ceramic tile can be used throughout the house to provide surfaces that are *(one wrong)*: a. durable; b. colorful; c. flexible; d. easy to clean.

2. All ceramic tile is made from pure clay. (T or F)

3. Tile is _____ at high temperatures to form a relatively hard material.

4. All tile has a glazed face. (T or F)

5. Nearly _____ million square feet of tile is installed each year in the United States.

6. Custom lots of ceramic tile are made by hand. (T or F)

7. Most ceramic tile is made in highly automated factories. (T or F)

8. The body of tile is called a _____.

9. Tile is fired in a _____ at high temperatures.

10. The temperature and length of firing determines the water permeability of the finished material. (T or F)

11. Highly permeable tiles are the least waterproof because they absorb more water. (T or F)

12. The most permeable tile is called _____ tile.

13. The least permeable tile is called _____ tile.

14. Tiles that will be exposed to water should be _____ permeable.

15. If a drop of water sits on top of the back of the tile that indicates the tile is towards the _____ side of the scale.

16. Wall tile is generally a _____ tile with a relatively soft glaze.

17. Wall tile is usually about _____ inch thick.

18. Tile that is specially shaped to form a border around a tile installation is called _____ tile.

19. Field tile is glazed on the top surface only. (T or F)

20. Paver tile is at least _____ inch thick.

21. Paver tile is intended for use on walls. (T or F)

22. Handmade, unglazed pavers are known as _____ tile or _____ tile.

23. Quarry tile is excellent for use on floors. (T or F)

24. Mosaic tile is always _____ inches or smaller.

(Continued on next page)

_____ 25. The lugs on lugged tiles automatically determine the proper spacing. (T or F)

_____ 26. Epoxy adhesives were developed in the 1970s. (T or F)

_____ 27. Adhesives should completely cover the back of the tile. (T or F)

_____ 28. Adhesives are applied with a _____ trowel.

_____ 29. Traditional mortar for ceramic tile is made from (one wrong): a. Portland cement; b. stone; c. sand; d. lime.

_____ 30. Dry-set mortar consists of a mixture of Portland cement, sand, and certain additives. (T or F)

_____ 31. Dry-set mortar mixed with latex or acrylic-modified liquids is weaker than dry-set mixed with water. (T or F)

_____ 32. Epoxy dry-set mortars can be used on the following surfaces (one wrong): a. plastic laminates; b. steel; c. copper; d. plywood.

_____ 33. Mastic is an organic adhesive that comes ready-mixed. (T or F)

_____ 34. Mastic is stronger than mortar. (T or F)

_____ 35. Tiles are attached to a substrate with _____ or mastic.

_____ 36. The space between tiles must be filled with a grout. (T or F)

_____ 37. Grout is a form of _____.

_____ 38. Grout comes in a wide array of colors. (T or F)

_____ 39. Tile grout comes in two forms — plain and _____.

_____ 40. Any kind of waterproof sheet material that is used between the tile installation and the substrate is called a _____.

_____ 41. Tar paper is a common waterproofing material used beneath ceramic tile. (T or F)

_____ 42. The abbreviation for chlorinated polyethylene is _____.

a. _____

b. _____

c. _____

d. _____

e. _____

f. _____

g. _____

43. Identify the following tools:

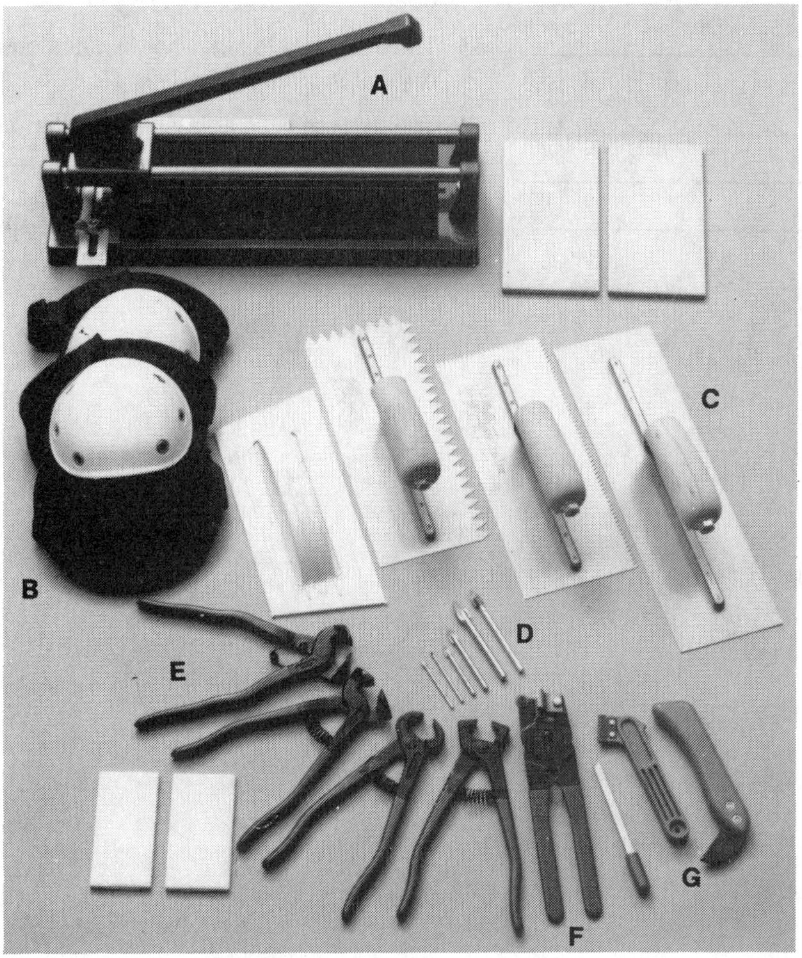

Fig. 48-1.

_____ 44. A wet saw is a small radial-arm saw fitted with a _____ blade.

_____ 45. There are many methods for installing tile. (T or F)

_____ 46. The proper methods and materials used in installing tile depend on the following factors (one wrong): a. structural loads expected; b. stiffness of the substrate; c. strength of tile; d. condition of substrate.

_____ 47. One factor that can affect the stiffness of the floor is the distance between supports. (T or F)

_____ 48. An isolation membrane reduces the chance that movement in the substrate will be transmitted to the tile. (T or F)

_____ 49. Thick-bed installation requires a mortar-setting bed that is _____ inch or _____ inches.

_____ 50. A thick-bed installation should be applied to metal lath. (T or F)

_____ 51. The mortar for a thick-bed installation should be applied in _____ layers.

(Continued on next page)

_____ 52. The layers include a scratch coat, a bed coat, and a _____ coat.

_____ 53. The adhesive for thin-set installations is made of _____ or dry-set mortar.

_____ 54. The thickness of adhesives for thin-set installations should be from _____ inch to _____ inch.

_____

_____ 55. A thin-set installation can be done over a concrete slab. (T or F)

To be used with Unit 52 **Name** ______________________

Pages 685-706 **Score: (32 possible)** ______

Study Unit 49
Stairs

1. The two main types of stairs are service and ____________.
2. The ____________ style is more expensive to construct.
3. A continuous-run stairway is straight. (T or F)
4. A stairway consisting of more than one run has a ____________ or winder at the angle.
5. Of the two angle treatments, ____________ is safer.
6. The part of a stair on which a person steps is called the ____________.
7. Treads are supported on each side by ____________.
8. Figure 49-1 shows a stairway with a ____________.

Fig. 49-1.

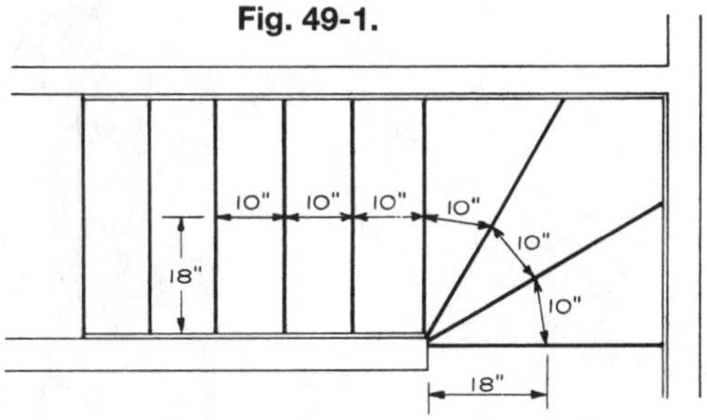

9. The section of the stairway from one step to the next is called a ____________.
10. The edge of a step projects out a little beyond the vertical section. This is called ____________.
11. The minimum amount of headroom required by FHA in designing a stairway is *(one right)*: a. 7′4″; b. 6′8″; c. 7′7″; d. 7′.
12. The full length of the vertical distance of a stairway is called the total ____________.
13. The full length of the stairway measured horizontally is called the total ____________.
14. The floor at the top of a stairway is called a ____________.
15. A landing is the same thing as a platform. (T or F)
16. The opening for a stairway requires extra framing members. (T or F)
17. The most important designing considerations for stairways are *(one wrong)*: a. width of the stairwell; b. amount of headroom to be allowed; c. whether it is open or closed on one side; d. relationship between riser height and tread width.
18. The vertical distance between one floor and the ____________ is total rise.
19. A stairway should be at least ____________ feet wide.

(Continued on next page)

a. _____

b. _____

c. _____

d. _____

e. _____

f. _____

g. _____

h. _____

20. Name the parts of the stairs shown in Fig. 49-2.

Fig. 49-2.

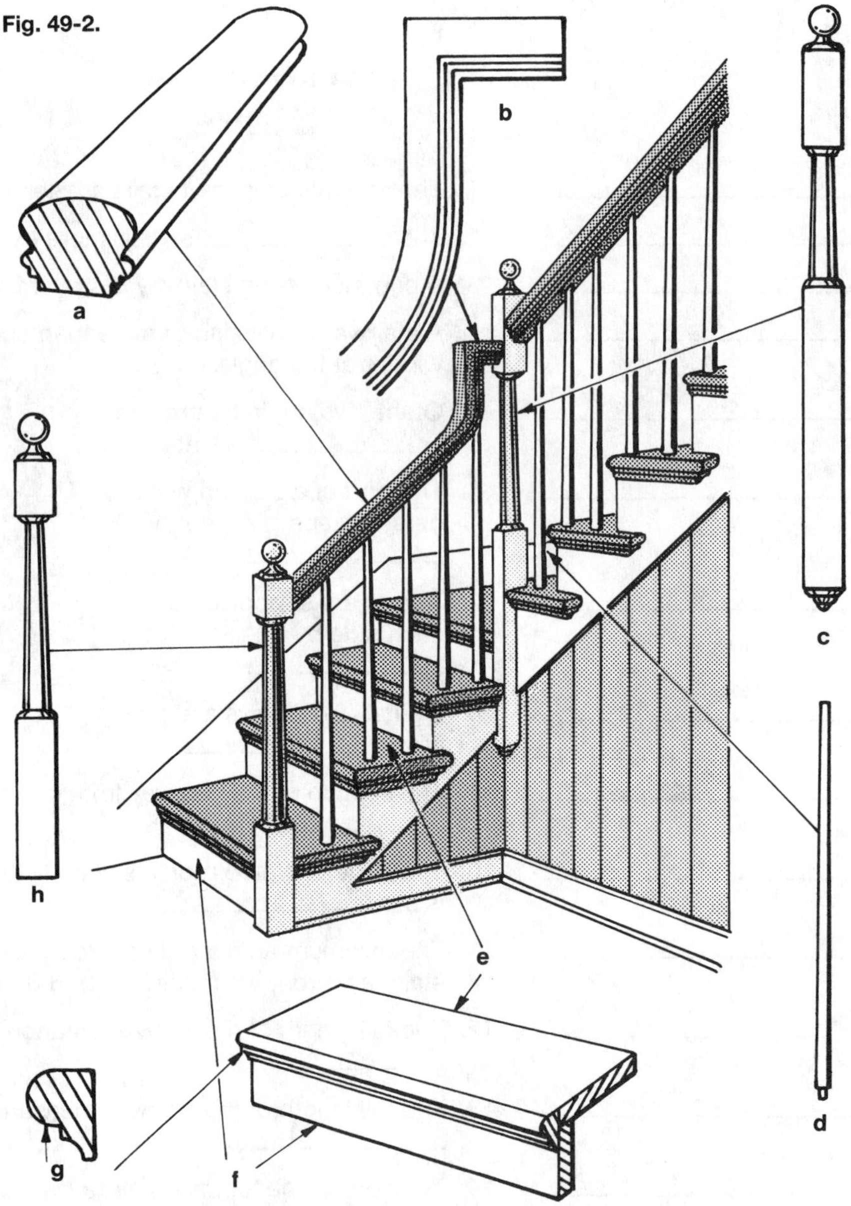

_____

21. A stairway designed with improper rise and _____ will be tiring to use.

_____

22. Since stringers are the first stairway parts to be erected, they must be (one wrong): a. made of hardwood; b. plumb; c. level; d. solidly fixed in place.

_____

23. _____ are not required for a basement stairway.

_____

24. Space between treads on a cleat stairway should be between 6 1/2 and _____ inches.

_____

25. To be sure the first tread on a cleat stairway is placed in proper position, begin by subtracting the thickness of the _____ from the determined riser height.

To be used with Unit 53

Pages 707-738

Name ______________________________

Score: (38 possible) ______

Study Unit 50
Cabinets and Built-Ins

a. ______________________
b. ______________________
c. ______________________
d. ______________________

1. Name the kinds of kitchen layouts shown in Fig. 50-1.

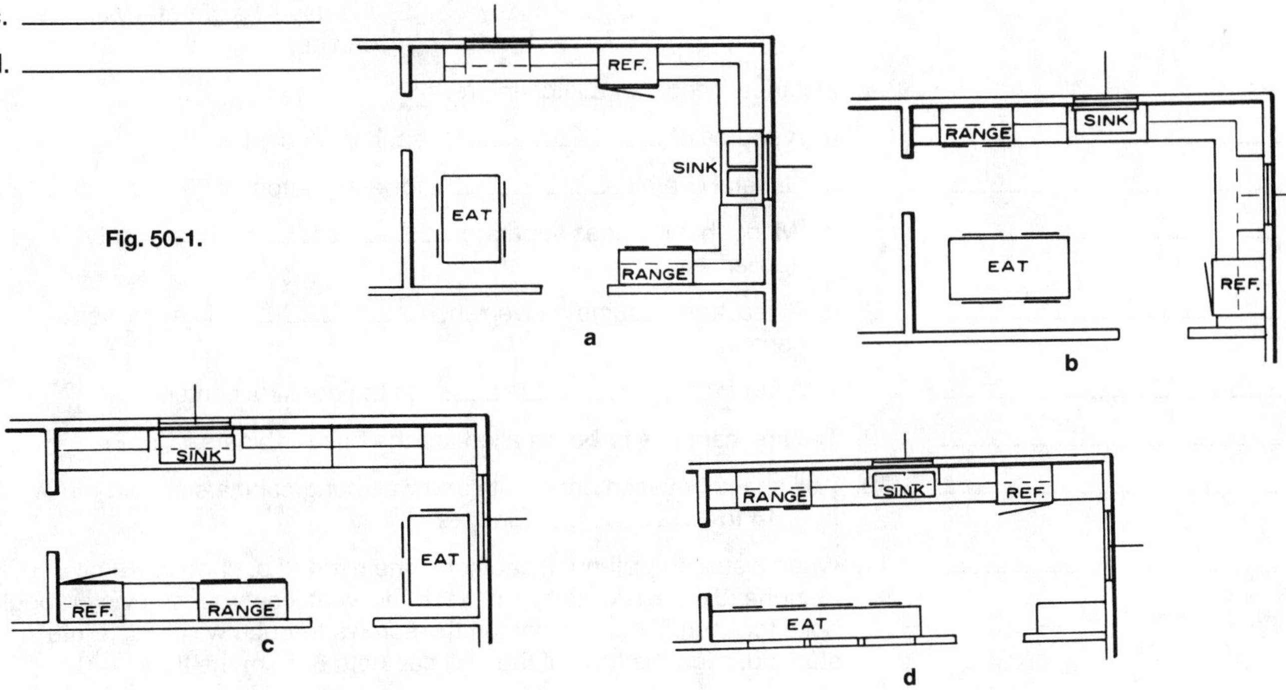

Fig. 50-1.

a. ______________________
b. ______________________
c. ______________________
d. ______________________

2. Complete these descriptions of kitchen layouts:
   a. Working units are located along one wall in the ______________ design.
   b. In the ______________ design, sink and stove are on one wall with the refrigerator on an adjoining wall.
   c. Working units in the ______________ design are located on opposite walls.
   d. The sink is on one end of the room and working units are on opposite walls adjoining the sink in the ______________ design.

____
3. The design usually found in small apartments is the ______________.

____
4. The main work centers of a kitchen are *(one wrong)*: a. food preparation; b. cooking; c. clean-up; d. dining.

____
5. Kitchen cabinets are produced in ______________ different ways.

____
6. When installing factory-built cabinets, measure along the wall ______________ inches from the floor.

(Continued on next page)

a. _____

b. _____

c. _____

d. _____

e. _____

f. _____

7. Match the dimensions at the left with the kitchen components at the right:

a. 24″
b. 60″
c. 12″
d. 21″
e. 36″
f. 30″

1. height of wall cabinets over a refrigerator
2. standard height of the counter of a base unit
3. width of the clean-up center
4. amount of counter space for serving center
5. height of wall cabinets over a sink
6. the usual height of a wall-hung cabinet

a. _____

b. _____

c. _____

d. _____

e. _____

8. Electrical needs for a kitchen are:

a. A source of _____ for each work center.

b. General overall _____ for entire kitchen.

c. Wiring that includes separate _____ for heavy-duty appliances.

d. At least one double convenience _____ in each work center.

e. A fan for proper _____ in the cooking center.

_____

9. The first cabinets to be installed are the base cabinets. (T or F)

_____

10. Wall space between base units and wall-hung cabinets should range from 14 to _____ inches.

_____

11. When installing wall-hung cabinets (one wrong): a. choose #9 or #10 roundhead screws or toggle bolts; b. allow at least four screws or toggle bolts for each cabinet; c. fasten the screws through wall and 1″ into studs; d. align the tops of the wall cabinets 6′ from the floor.

_____

12. Kitchens usually consist of cabinets that are (one wrong): a. built at the factory ready for installation; b. purchased in parts and sections ready for the carpenter to install on the job; c. built piece by piece, following the architect's plans; d. built individually to the carpenter's own specifications.

_____

13. Bathroom cabinets are similar in construction and installation to kitchen cabinets. (T or F)

_____

14. As a rule, bathroom cabinets are a little larger than kitchen cabinets. (T or F)

_____

15. Countertops are usually covered with plastic _____.

_____

16. The material most often used for countertop coverings can be described as (one wrong): a. thin; b. requiring special tools; c. brittle; d. available in many sizes.

_____

17. Laminates must be fastened to a core _____ inch thick.

_____

18. Good cores include (one wrong): a. solid wood; b. plywood; c. hardboard; d. particleboard.

_____

19. The core is fastened to the outer covering with _____.

_____ 20. A drawer with one slide can be mounted in a frameless cabinet. (T or F)

_____ 21. Double slides are available that allow the drawer to extend fully away from the cabinet. (T or F)

_____ 22. European-style hinges are designed to be used on frameless cabinets. (T or F)

_____ 23. The approximate time needed to install a base cabinet is _____ hour.

Study Unit 51
Interior Trim and Interior Doors

_____

1. The paint or natural finish to be used on interior trim determines the species of wood to select. (T or F)

_____

2. The moisture content for interior trim should be from 6 to _____ percent.

_____

3. Ponderosa pine would be a good wood to use when the trim is subject to hard usage. (T or F)

4. Before interior trim can be installed, the following construction must be concluded:

a. _____

a. The finish floor must be _____.

b. _____

b. The surfaces of walls and floors should be scraped _____.

c. _____

c. The location of all _____ should be clearly marked.

_____

5. Door and window frame trimming should be installed before base and wall moldings. (T or F)

_____

6. Openings in walls for insertion of interior doors are usually framed with two- or three-piece jambs. (T or F)

_____

7. It is possible to purchase a complete installation, including the jamb, with the door fitted and prehung. (T or F)

_____

8. The standard interior door thickness is (one wrong): a. 1 1/4"; b. 1 1/2"; c. 1 3/8"; d. 1 3/4".

_____

9. Folding and sliding doors are usually thicker than standard interior doors. (T or F)

a. _____

b. _____

c. _____

d. _____

10. Identify the doors shown in Fig. 51-1.

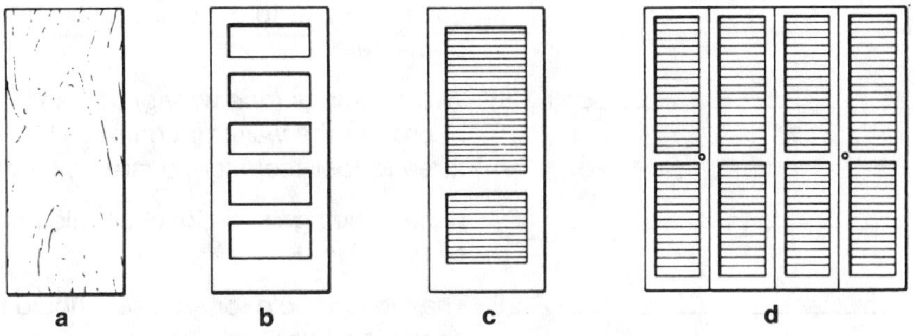

Fig. 51-1.

a   b   c   d

_____

11. The door frequently chosen for closets is the _____ door.

12. Match the interior door widths at the left with the door positions at the right:

a. _____

a. 2'   1. closets

b. _____

b. 2'6"   2. bathrooms

c. _____

c. 2'4"   3. bedrooms

d. _____

d. 6'   4. openings for sliding doors

(Continued on next page)

_____

_____

_____

_____

_____

_____

a. _____

b. _____

c. _____

d. _____

_____

_____

_____

_____

_____

_____

_____

_____

13. The standard interior door height is 7' 2". (T or F)

14. Standards for the direction of door swing are *(one wrong)*: a. against a blank wall; b. into a hallway; c. in the direction a person would naturally enter a room; d. so that the door doesn't interfere with the swing of another door.

15. The hinges for an interior door should be loose-pin _____ hinges.

16. A door should always be hung with three hinges. (T or F)

17. The door that allows greatest accessibility to a closet is *(one right)*: a. sliding; b. door that swings into the room; c. sliding bypass door; d. bifold door.

18. A good choice for a place where the door is seldom closed is the sliding pocket door. (T or F)

19. Match the window trim parts at the left with their descriptions at the right:

a. casing
b. sash stop
c. stool
d. apron

1. horizontal trim member that laps the window sill
2. finish member below stool
3. trim around the upper part of a window
4. allows the sash to move freely

20. Window trim is always made of wood. (T or F)

21. The first piece of window trim to be installed is the _____.

22. The apron should extend the width of the _____.

23. The last trim to be installed in a room is the _____ molding.

24. Base molding can be purchased in a great many sizes and shapes. (T or F)

25. Ceiling moldings are used *(one wrong)*: a. for decoration; b. as structural members which help support the ceiling; c. as junctions between wall and ceiling; d. to conceal any uneven appearance of paneling, drywall, or plaster.

26. Wall moldings *(one wrong)*: a. can break up two different wall finishes; b. protect the wall from damage by chair backs; c. are installed from three to four feet high; d. are no longer used.

27. Base moldings in a closet should match those in the room the closet serves. (T or F)

28. The clothes pole for a closet should be about _____ inches above the floor.

29. The clothes pole fits into holes cut in two strips of molding fastened to the sides of the closet. (T or F)

30. The closet shelf rests on top of the _____ strips.

Study Unit 52
Chimneys and Fireplaces

_____ 1. A chimney must be of masonry construction. (T or F)

_____ 2. A chimney performs the following (one wrong): a. produces a draft; b. helps supply fresh air to the fire; c. expels smoke and harmful gases; d. heats the house.

_____ 3. A chimney produces a better draft when outside temperatures are low. (T or F)

_____ 4. A chimney situated in the interior area of a house has a better draft than one on an outside wall. (T or F)

_____ 5. The gases and smoke go up a passage in the chimney called a _____.

_____ 6. The two things most important to the efficiency of a chimney are flue size and _____ height.

_____ 7. A chimney will not draw properly unless the topmost part of the chimney is (one wrong): a. 2' above a roof with a ridge; b. 3' above a flat roof; c. even with any raised part of a roof within 10' of the chimney; d. fitted with a hood to control irregular air currents.

_____ 8. A chimney is usually the heaviest single part of a house. (T or F)

_____ 9. Mortar for laying chimney flue brickwork should be made with _____.

_____ 10. _____ mortar does not resist heat and flue gases well.

_____ 11. A house with one heating unit and a fireplace requires _____ flues.

_____ 12. A fireplace must have its own flue. (T or F)

_____ 13. A chimney flue ought to be lined because (one wrong): a. the lining protects mortar and bricks in the chimney; b. the lining prevents cracks from developing in the chimney; c. this cuts the cost of construction; d. lined flues are safer and more efficient.

_____ 14. Flues should rise upward at no greater angle than (one right): a. 40 degrees; b. 45 degrees; c. 30 degrees; d. 35 degrees.

_____ 15. An angle of _____ degrees or less is the best rise for a flue.

_____ 16. Flues must always be built of solid masonry. (T or F)

_____ 17. Fireplaces, flues, chimneys and similar construction components must abide by the recommendations of an organization called _____.

_____ 18. In a chimney containing more than one flue, 3 3/4 to 4 inches should be allowed for separating them. (T or F)

_____ 19. Fireplaces and stoves are connected to flues by smoke pipes. (T or F)

(Continued on next page)

20. Woodwork should never be closer than _____ inches to smoke pipes.

21. When a smoke pipe must pass through wood to reach a flue, one of the following precautions should be taken:

    a. _____

    a. Install at least 4" of incombustible material or _____ around the pipe.

    b. _____

    b. Surround the pipe with a shield at least _____ inches larger than the pipe.

_____

22. Between chimney walls and the wood members of the house, a _____ inch space should be allowed.

23. Complete the following statements on chimney-top construction:

    a. _____

    a. Moisture is prevented from entering between flue lining and brickwork by a concrete _____.

    b. _____

    b. The _____ lining extends 4" above this.

    c. _____

    c. Rain and downdraft are kept out of the chimney by _____.

    d. _____

    d. The hazard of sparks can be controlled by means of spark _____.

_____

24. The depth of a fireplace has little to do with good draft. (T or F)

_____

25. Fireplace construction has been standardized for maximum safety and function. (T or F)

_____

26. Floor framing around a fireplace must be stronger than in other areas. (T or F)

27. Match the fireplace parts at the left to the descriptions at the right:

    a. _____

    b. _____

    c. _____

    d. _____

    e. _____

    f. _____

    g. _____

    h. _____

    | a. jambs | 1. prevents downdraft |
    | b. hearth | 2. the "floor" of the fireplace |
    | c. walls | 3. exposed sides of fireplace |
    | d. throat | 4. back and sides of fireplace |
    | e. lintels | 5. extends from the top of the throat to the bottom of the flue |
    | f. smoke shelf | 6. goes across the top of the opening to support the masonry |
    | g. damper | 7. regulates throat opening |
    | h. smoke chamber | 8. widest part of the flue |

_____

28. Modified fireplaces are made in a _____.

_____

29. A modified fireplace (one wrong): a. produces less heat than a conventional fireplace; b. contains all necessary fireplace parts; c. reduces chances of faulty construction; d. is less likely to smoke than a conventional fireplace.

_____

30. A modified fireplace requires the construction of a conventional chimney. (T or F)

_____

31. A complete fireplace plus chimney, called a _____ unit, can be purchased from the factory.

Name _____

Study Unit 53
Protection against Decay and Insect Damage

_____

1. Wood decay is caused by *(one wrong)*: a. extreme conditions of dryness; b. extreme conditions of wetness; c. water that works itself into wood parts of a house and soaks into the wood; d. decay organisms that grow in damp wood.

_____

2. The two kinds of termites are dry-wood and _____.

_____

3. The kind of termite hardest to control is the _____.

_____

4. The termite that does the most serious kind of damage is the _____.

_____

5. Decay in wood can easily be seen on its surface. (T or F)

_____

6. Wood decay develops most rapidly in temperatures of 70 to _____ degrees F.

_____

7. Wood decay organisms *(one wrong)*: a. are killed by kiln-drying of lumber; b. cannot grow in dry wood; c. cause wood to change in appearance and properties; d. are killed by cold temperatures.

_____

8. Sapwood is more subject to decay than heartwood. (T or F)

_____

9. Another name for wood decay is dry rot. (T or F)

10. Complete the following statements on prevention of decay:

a. _____

a. Construction lumber should be fully seasoned, not _____.

b. _____

b. Wood should not be fully enclosed or _____ until it is thoroughly dry.

c. _____

c. Make sure there is proper clearance between wood construction members and the _____.

d. _____

d. Such units as steps, wall plates, and posts should be insulated with concrete or _____.

e. _____

e. Runoff of _____ should be provided for on such sections as roofs.

f. _____

f. _____ should be installed around chimneys, doors, and windows.

g. _____

g. A house should be fitted with _____ and downspouts.

_____

11. There are several kinds of _____ that can be administered to wood to preserve it.

_____

12. Preservatives can be administered *(one wrong)*: a. by pressure treatment; b. by painting; c. by soaking; d. by spraying.

(Continued on next page)

13. Water vapor given off by indoor activities can cause wood decay unless the following precautions are taken (one wrong): a. installation of proper vapor barrier on warm side of walls; b. good ventilation in attic spaces; c. leaking pipes are ignored; d. clothes washing done in basement.

_____

14. Using dry lumber in construction is the simplest way to prevent wood decay. (T or F)

_____

15. The termites that bore from the ground through tunnels into wood are the _____ termites.

_____

16. _____ termites fly directly to wood.

_____

17. Timber, woodwork, and furniture that are finished or painted are less likely to be attacked by termites than raw wood. (T or F)

_____

18. Protection against termites should be taken into consideration during both planning and _____.

_____

19. Soil under the house should be kept as _____ as possible.

_____

20. The best foundation for protection against ground termites is properly reinforced _____.

_____

21. The best protection against ground termites is to treat the _____ around and under foundations.

_____

22. Any wood used in decorative fences and gates should be pressure-treated with a good _____.

23. The following precautions should be taken to prevent damage from dry-wood termites:

a. _____

a. Give the exterior of the house several coats of _____.

b. _____

b. Fill all cracks and crevices with mastic _____ or plastic wood.

c. _____

c. Carefully inspect all _____.

d. _____

d. Put _____ on all outside doors and windows.

e. _____

e. Use construction lumber that has been given a _____ treatment.

_____

24. When using pesticides, follow the label _____ carefully.

_____

25. Carpenter ants eat wood. (T or F)

_____

26. To prevent carpenter ants, the chemical treatment must be applied to the _____.

_____

27. The most common type of beetle is the _____ beetle.

_____

28. The second type of beetle is the _____ beetle.

_____

29. The greatest beetle activity is in wood with a moisture content between _____ and 20 percent.

_____

30. All wood and wood products are subject to decay. (T or F)

_____

31. Dry-wood termites are found principally in (one wrong): a. Florida; b. Maine; c. southern California; d. Hawaii.

_____

32. Fungi require air, warmth, _____ and moisture to grow.

_____ 33. Fungi grows most rapidly at temperatures of about 70 to
_____ degrees.

_____ 34. Fungi cannot grow in wood with a moisture content of _____
percent or less.

_____ 35. Foundation walls should have a clearance of at least _____"
above the exterior grade.

_____ 36. Water-repellent preservatives must have a minimum of _____
percent by weight of pentacholorophenal.

Study Unit 54
Scheduling

_____

1. The responsibilities of the general contractor include *(one wrong)*: a. arranging for materials needed by the subcontractors; b. job scheduling; c. material scheduling; d. coordinating construction work so it functions smoothly.

_____

2. Successful general contractors usually are former _____.

_____

3. A general contractor should hire as few subcontractors as possible. (T or F)

_____

4. Some of the jobs turned over to subcontractors include *(one wrong)*: a. plumbing; b. job scheduling; c. concrete and masonry; d. painting and decorating.

_____

5. The delivery of needed materials to the building site depends on *(one wrong)*: a. the date set for completion of the project; b. the size of the project and its type; c. the climate; d. the size of the work force.

_____

6. The lumber supplier's responsibilities include *(one wrong)*: a. sending additional or missing items to the building site; b. accepting leftover materials for credit; c. keeping a financial record of the cost of the materials; d. collecting money from the contractor for payment for materials.

_____

7. Filling earth around the outside of a foundation wall is called _____.

_____

8. Filling earth around the foundation cannot be done until and unless *(one wrong)*: a. the wood-frame walls of the building have been framed; b. the foundation walls have been braced; c. the framing has progressed far enough to provide sufficient weight to the foundation; d. exterior walls of the foundation have been moisture-proofed.

(Continued on next page)

a. _____

b. _____

c. _____

d. _____

e. _____

f. _____

g. _____

h. _____

i. _____

j. _____

k. _____

l. _____

m. _____

n. _____

_____

_____

_____

_____

_____

9. Match the steps in home construction at the left with the workers who do them at the right. Some numbers will be used more than once:

a. installing pipelines in subsoil
b. building chimney and/or fireplace
c. deciding when to backfill
d. finishing interior after plaster is dry
e. planting lawns and shrubbery
f. digging the basement
g. treating for termites
h. installation of doors and windows
i. installing electrical wiring, switches, and fixtures
j. framing walls, floors, and roof
k. putting the final finish on trim, doors, woodwork, and other similar jobs
l. installing fixtures in bathroom
m. doing the final cleanup
n. putting in driveways and sidewalks

1. carpenter
2. excavator
3. plumber
4. general contractor
5. painter
6. concrete worker
7. masonry worker
8. landscaper
9. termite control specialist
10. electrician

10. Interior trim is added immediately after walls are plastered or covered with drywall. (T or F)

11. A building is inspected only once, in its final stages. (T or F)

12. The contractor makes a _____ list of things that the owner wants corrected.

13. The easiest way to keep track of scheduling is with a _____ chart.

14. A scheduling method that shows the interrelationships between various tasks is called the _____ _____ method.

Study Unit 55
Plumbing, Electrical, and HVAC Systems

_____

1. Plumbers must have skills in *(one wrong)*: a. woodworking; b. metalworking; c. painting; d. welding.

_____

2. The letters used to identify the Uniform Plumbing Code are _____.

_____

3. Floor plans, elevations, and specifications for plumbing are developed by an _____.

_____

4. Residential plumbing consists of the _____ supply system and the _____ disposal system.

_____

5. A water meter is always located inside the house. (T or F)

_____

6. If a public water system is not available, the water must come from an underground _____.

_____

7. In a plumbing system, the _____ must go through the roof.

_____

8. Framing for bathtubs requires additional joists for support. (T or F)

_____

9. Prefabricated showers and tubs are made of _____.

_____

10. Solder of lead and tin can be used to seal joints on copper tubing. (T or F)

_____

11. The Drinking Water Act became law in _____.

_____

12. The letters used to identify polybutylene are _____.

_____

13. A nonlead solder is a combination of tin, copper, and _____.

_____

14. NCE stands for _____ _____ _____.

15. An electrician must have a _____ to install house wiring.

16. Identify these electrical symbols:

a. _____   a.   S

b. _____   b.   S₃

c. _____   c.   ○ or ⊖

d. _____   d.   ○PS or ⊖PS

e. _____   e.   ⊖

f. _____   f.   ⊖R

_____

17. All power comes into the building through the _____ entrance wires.

(Continued on next page)

_____

_____

_____

_____

_____

_____

_____

_____

_____

_____

_____

_____

_____

_____

_____

_____

_____

_____

_____

_____

_____

_____

_____

_____

_____

_____

_____

18. The following are the basic kinds of circuits (one wrong): a. appliance; b. general purpose; c. general; d. special purpose.

19. The simplest and least expensive wiring is called _____ _____ _____.

20. The bare copper wire in a circuit is used for _____.

21. Another name for nonmetallic sheathed cable is _____.

22. Another name for outlet boxes is _____ boxes.

23. All outlet boxes are made of metal. (T or F)

24. There are two types of metal conduit, namely thin and _____.

25. Wiring is done in two stages, namely the _____ and the finish.

26. The electrical meter is always installed inside the house. (T or F)

27. Inspectors must approve wiring at least _____ times.

28. Research is making heating and cooling systems more _____-efficient.

29. The four letters used to describe a heating, ventilating, and air-conditioning system are _____.

30. Heating, ventilation, and air-conditioning needs vary from region to region in the following ways (one wrong): a. local conditions; b. climate; c. federal regulations; d. availability of resources.

31. Electricity is more expensive in the Northwest than oil. (T or F)

32. The Northwest gets much of its electricity from _____ sources.

33. With a forced hot-air system, a _____ in the furnace circulates the warm air through ducts and registers.

34. Forced hot-air systems respond slowly to outdoor temperature changes. (T or F)

35. A hot-air system cannot be used in homes without basements. (T or F)

36. A heat pump is a device that can only heat the air in a house. (T or F)

37. A heat pump is connected to standard duct systems. (T or F)

38. Filters should be used in a hot-air system. (T or F)

39. Air cleaners that remove pollen, fine dust, and other irritants are _____.

40. A humidifier removes moisture from the air. (T or F)

41. Ducts are made of sheet metal, rigid fiberglass, or flexible fiberglass _____.

42. Outlets for heating should be located high on the walls. (T or F)

43. Warm air sinks towards the floor as it cools. (T or F)

44. When a home has a basement, ducts are run above the ceiling joists and _____ the floor joists.

45. Homes without basements often have the ducts located within the slab. (T or F)

_____ 46. Hot water and steam for a heating system are generated in a
_____.

_____ 47. Two devices used to transfer heat into a room used with hot water or steam heat are radiators or _____.

_____ 48. Radiant heating systems can combine elements of hot-air and _____ systems.

_____ 49. In a radiant system, heat is transferred to a material that then _____ the heat to people.

_____ 50. With a radiant heating system, the heating coils or cables are buried within the ceiling, _____, or walls.

_____ 51. With a radiant hot-water system, heated water is circulated through tubing embedded in a _____ floor.

_____ 52. Heat recovery ventilators are sometimes called air-to-air _____ exchanges or _____ recovery ventilators.

_____ 53. A heat recovery ventilation system usually is part of the regular heating system. (T or F)

_____ 54. Whole-house ventilation can be accomplished with a large fan located in the highest ceiling in the house. (T or F)

_____ 55. A two-pipe hydronic system uses one pipe to carry heated water to the room and one pipe to return the cooled water to the boiler. (T or F)

_____ 56. Hot-air systems are designed to keep people comfortable, while radiant systems heat air. (T or F)

_____ 57. HRV stands for heat recovery _____.

_____ 58. A device that can heat or cool the air in a house and that is connected to standard duct systems is known as a _____ _____.

Study Unit 56
Interior and Exterior Painting

_____

1. Proper application of paint can prolong the life of a home and improve its _____.

2. Interior paints have the following purposes:

a. _____
    a. They make surfaces easy to _____.

b. _____
    b. They impart _____ resistance.

c. _____
    c. They seal surfaces from _____.

d. _____
    d. They enhance a desired _____ effect.

_____

3. The most popular interior paint is *(one right)*: a. high-gloss enamel; b. flat alkyd; c. latex; d. semigloss enamel.

_____

4. Drywall and plaster surfaces should be treated with sizing before painting. (T or F)

_____

5. The best paint for bathroom or kitchen is semigloss _____.

6. Interior wall finishing can be made a more pleasant job if the following pointers are kept in mind:

a. _____
    a. Choose the best _____ for the job.

b. _____
    b. Apply it according to the directions shown on the _____.

c. _____
    c. Use _____ or rollers of high quality.

d. _____
    d. Make sure walls are properly _____.

e. _____
    e. Immediately wipe up all spills or _____.

f. _____
    f. Make sure the room is _____ and well ventilated.

g. _____
    g. Tools must be _____ immediately after use.

_____

7. Floors and furniture should be protected with newspaper or a _____ cloth.

_____

8. All cracks and nail _____ should be filled before painting.

_____

9. If a crack is large, fill it with _____ plaster.

_____

10. It is not necessary to allow patch repairs to dry before painting. (T or F)

_____

11. A professional paint job can be done without removing such hardware as switch plates and door knobs. (T or F)

_____

12. A narrow strip of paint should be applied around windows and doors and along _____.

_____

13. Applying a narrow strip of paint around the edge of a wall is called _____ in.

_____

14. One entire wall should be finished before going to the next. (T or F)

(Continued on next page)

_____   15. A roller is easier to use than a brush. (T or F)

_____   16. If a brush is used, the tip should be dipped one-half the length of the bristles. (T or F)

_____   17. Paint should be applied to a wall from the top downward. (T or F)

_____   18. Follow these techniques for painting woodwork (one wrong): a. wait until walls are dry; b. use a wide brush; c. work as quickly as possible; d. avoid touch-up on areas that have dried.

_____   19. Painting window sash can be made easier by applying _____ to the glass.

_____   20. There is no set procedure for painting doors. (T or F)

_____   21. Old paint can be removed by scraping or with a heat _____.

_____   22. Protecting windows and other areas from being splattered with paint is called _____ _____.

_____   23. Interior paints can be applied with spray guns, _____, and brushed.

_____   24. Exterior painting should be done before doors and windows are installed. (T or F)

_____   25. Oil-based paints are more popular than water-based paints. (T or F)

_____   26. Oil-based paints usually contain _____ oxide.

_____   27. Paint protects the exterior of a home. (T or F)

_____   28. Paint's most important function is to keep excess _____ out of the wood.

_____   29. Paint keeps exterior siding from cupping and warping. (T or F)

_____   30. Latex paint has the following characteristics (one wrong): a. fast drying; b. easy to apply; c. can be applied to a damp surface; d. can be applied at any temperature.

_____   31. When applying a three-coat system, the first coat should be primer or _____.

                              32. Some things to remember when painting are:

a. _____   a. Seal knots and pitch spots with _____.

b. _____   b. Remove old scaling and _____ paint.

c. _____   c. Rough up slick, shiny surfaces with _____ or a wire brush.

d. _____   d. Sink any _____ below the surface.

e. _____   e. Clean all gutters and _____.

f. _____   f. Remove all loose or dry _____ from around windows.

_____   33. Moisture condensation has a great effect on exterior paint surfaces. (T or F)

_____   34. Exterior walls must be protected against moisture from both inside and outside surfaces. (T or F)

_____   35. Outside wall paint is protected from moisture that comes from inside a home by a _____ barrier.

144

Study Unit 57
Building Architectural Models

_____   1. Building a model converts a drawing into a _____-dimensional object.

_____   2. Architectural models of homes are usually built to a scale of _____" = 1'.

_____   3. Architectural models of large buildings are built to a scale of _____" = 1'.

_____   4. Shrubbery can be represented by _____ .

_____   5. There are two basic types of building models, namely, architectural and _____.

_____   6. The building model that shows primarily what the exterior of a building will be is the _____ model.

_____   7. A _____ model is an actual construction of the building to scale.

_____   8. A house model is best built to a scale of *(one right)*: a. one-half; b. one-fourth; c. one-eighth; d. one-sixteenth.

_____   9. Woods used for a scale model should be hardwoods. (T or F)

_____   10. The base of a model can be made of *(one wrong)*: a. hardboard; b. cement slab; c. plywood; d. particleboard.

_____   11. It is best to assemble the model with regular-size nails. (T or F)

_____   12. Building a model is merely an exercise in the use of tools. (T or F)

_____   13. The roof of a model should be constructed as a single unit so it can be removed. (T or F)

_____   14. Shrubbery for a model house can be made from _____.

_____   15. Lumber for structural members should be made of *(one wrong)*: a. basswood; b. ash; c. yellow poplar; d. redwood.

_____   16. Attach shingles to the model with _____" nails or staples.

_____   17. Model-size trusses can be constructed using a _____.

_____   18. Termite shields for a house should be made of thin aluminum or _____ foil.

_____   19. A model house without a basement can use a piece of _____" plywood to represent the concrete slab.

(Continued on next page)

_____ 20. A beam made of layers of thin wood should be painted _____ to represent a steel I-beam.

_____ 21. Wood siding can be made from thin _____ wood.

_____ 22. A thin veneer can be used for roof _____.

_____ 23. It is easy to build miniature parts for millwork. (T or F)

Study Unit 58
Construction Systems

_____

_____

_____

_____

_____

1. A wood pile is a round or rectangular post driven into the soil. (T or F)

2. A technique known as _____ involves using a high-pressure stream of water to install a pipe.

3. A pile driver is the best way of putting a pile into the ground. (T or F)

4. Using 2 × 6s instead of 2 × 4s as wall studs permits a builder to install insulation that is _____ percent thicker.

5. The mechanical fasteners shown in Fig. 58-1 are known as _____ clips.

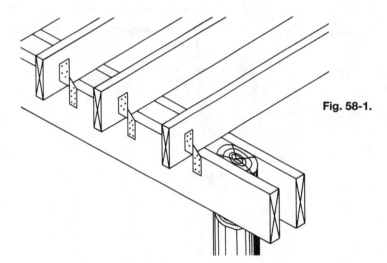

Fig. 58-1.

_____

_____

_____

_____

_____

_____

6. Fire _____ block the flow of air upward inside walls.

7. Materials are rated for fire resistance by the *(one right)*: a. American Association of Fire Chiefs; b. Underwriters' Laboratories; c. Associated General Contractors of America; d. United Brotherhood of Carpenters and Joiners.

8. Using a mineral wool insulation in a wall decreases the wall's fire resistance. (T or F)

9. Type-X gypsum board is designed specifically to resist sound. (T or F)

10. Metal studs do not transmit sound as well as wood studs. (T or F)

11. The permanent wood foundation *(one right)*: a. is difficult to install; b. can be installed only in good weather; c. can be built of untreated lumber and plywood; d. can be partially built in a factory.

(Continued on next page)

_____ 12. Permanent wood foundation costs more than a comparable foundation of poured concrete or concrete block. (T or F)

_____ 13. Fasteners used for a permanent wood foundation can be of plain steel. (T or F)

_____ 14. Those areas of the earth that are subject to earthquake activity are known as _____ zones.

_____ 15. In earthquake-resistant construction, each successive structural member above the foundation is tied to it. (T or F)

_____ 16. Wood connectors are used to reinforce all structural joints. (T or F)

_____ 17. The following sections of the country are subject to earthquakes _(one wrong)_: a. West Coast; b. Hawaii; c. Florida; d. East Coast.

_____ 18. Truss frames are placed 16″ on center. (T or F)

_____ 19. Truss frames are built on site. (T or F)

_____ 20. Truss frames for an entire house can be erected in less than _____ days.

_____ 21. The truss-framed system can be used only for single-story houses. (T or F)

Study Unit 59
Remodeling and Renovation

_____

_____

_____

_____

_____

_____

1. Paint peeling from random areas of a house's exterior is often caused by _____ from inside the house.

2. A swayback or leaning roof ridge on an older home can indicate _____ settling.

3. During the winter, wood window frames retain heat inside a house better than aluminum frames. (T or F)

4. An attic should have at least _____ gable vent(s).

5. The interior walls of most older homes are made of drywall. (T or F)

6. An additional room often may be added to an existing house in the _____, the basement, or the garage.

7. Match each remodeling job to the percentage of its cost by which the value of the house increases:

a. _____
b. _____
c. _____
d. _____

   a. 25-30%   1. kitchen remodeling
   b. 50%   2. adding a full bath
   c. 80-100%   3. replacing windows and doors
   d. 75-80%   4. adding insulation

_____

8. The three steps in adding a room to an existing house are research, _____, and construction.

_____

9. The major difference between new construction and remodeling is the _____ process.

_____

10. Windows in older homes are held in place by nails driven into the exterior casing. (T or F)

_____

11. Some prehung doors are designed to fit into existing door frames. (T or F)

_____

12. Doors less than 7' high need only two hinges. (T or F)

_____

13. Exterior steel doors are made with a(n) _____ core.

_____

14. All outside walls are bearing walls. (T or F)

_____

15. The simplest way to locate studs in a wall is with a stud finder. (T or F)

_____

16. Before cutting the opening for a new door in an existing wall, cut a(n) _____ opening.

_____

17. Partitions parallel to floor and ceiling joists are usually load-bearing. (T or F)

_____

18. A _____ is a prop placed against or beneath an object to support it.

(Continued on next page)

_____   19. A _____ is a heavy wood beam placed across the top of several shores as a horizontal support.

_____   20. A leak in a roof will usually show up on the ceiling as a wet spot directly below the leak. (T or F)

_____   21. If a single wood shingle has a wide crack, it is best to *(one right)*: a. repair it using a sheet metal patch; b. replace the shingle; c. patch it with an asphalt shingle; d. cover the crack with roofing paper.

_____   22. Nails, rather than screws, should be used to straighten warped siding boards. (T or F)

_____   23. A suspended ceiling reduces noise in four ways. (T or F)

_____   24. A suspended ceiling absorbs a large amount of the noise striking its surface. (T or F)

_____   25. The grid system for a suspended ceiling should be at least _____" to 2 1/2" below the bottom of the framing.

_____   26. In most cases, the main runners for a suspended ceiling should be installed parallel to the ceiling joists. (T or F)

_____   27. Tile should be cut with the face down. (T or F)

_____   28. Standard ceiling fixtures and _____ may be used with a suspended ceiling.

_____   29. The panels used under ceiling lighting must be _____.

_____   30. When installing ceiling panels, keep your fingers off the finished side of the board. (T or F)

Study Unit 60
Manufactured Housing

_____

1. Houses built stick by stick on a lot require more skilled personnel than factory-built structures. (T or F)

_____

2. Another name for factory-built housing is _____ housing.

_____

3. The first parts of houses to be prefabricated were *(one right)*: a. beams; b. trusses; c. flooring; d. roof girders.

_____

4. A method of house construction developed to promote industrialized housing is called the _____ Method.

_____

5. In the method of house construction developed by the National Lumber Manufacturers Association, the basic unit of measurement is *(one right)*: a. 4″; b. 6″; c. 12″; d. 16″.

_____

6. A prefabricated house consists of only the _____ of the house.

_____

7. About one-third of the total cost of a prefabricated house is for the _____.

8. A sectional home can be described as:

a. _____

   a. Completely finished on the _____.

b. _____

   b. Built on an _____ line.

c. _____

   c. A complete home built in _____.

d. _____

   d. Moved to the building site and then _____.

_____

9. A factory-built building unit designed to be used by itself or added to other units is a _____.

10. Match the manufactured housing at the left with the definitions at the right:

a. _____

b. _____

c. _____

d. _____

e. _____

f. _____

   a. recreational vehicle
   b. mobile home
   c. expandable mobile home
   d. sectional home
   e. double-width mobile home
   f. modular unit

1. made of two or more units joined to make a single house
2. a dwelling unit built for movement from one place to another
3. built on a chassis and meant to be used as a permanent dwelling
4. factory-built units for use by themselves or with similar units to make larger structures
5. a mobile home with parts that can be collapsed for transporting
6. a two-section mobile home, each with its own chassis

(Continued on next page)

_____ 11. The maximum width of mobile homes is *(one right)*: a. 10′; b. 14′; c. 12′; d. limited by what each state will allow on the highway.

_____ 12. The only manufactured housing that does not have to be moved on a truck is the recreational vehicle. (T or F)

_____ 13. Because lighter materials are generally used for framing, _____ homes differ from standard housing.

_____ 14. Workers in factories building mobile homes must be highly skilled. (T or F)

_____ 15. Mobile homes are stronger in construction than modular housing. (T or F)

_____ 16. A custom-built house is stronger than a modular home. (T or F)

_____ 17. Factories producing modular houses require much heavy-duty equipment. (T or F)

_____ 18. Modular homes are built of standard-size building materials. (T or F)

_____ 19. The use of various types of assembly processes to manufacture houses is called _____ housing.

_____ 20. Kitchens and bathrooms are the most complicated rooms in a house. (T or F)

NOTES

NOTES

NOTES